THE HOOPHOUSE HANDBOOK

SECOND EDITION

EDITED BY LYNN BYCZYNSKI
GROWING FOR MARKET

ISBN 978-0-9779781-9-9

Fairplain Publications Inc.
Growing for Market
PO Box 3747
Lawrence, Kansas 66046

785-748-0605
www.growingformarket.com

On the cover: Lettuce grown in Haygrove Tunnels at the Kansas State Research and Extension Center in Olathe, Kansas.

About this book

The Hoophouse Handbook - Second Edition includes articles published in the first edition (2003), in the Hoophouse Update (2011), and in recent issues of *Growing for Market* magazine. Articles without a byline were written by Lynn Byczynski, the editor of *Growing for Market*. Other articles were written by experienced hoophouse growers from across the United States. Biographical and contact information for them appears at the end of their articles.

Photographs without photo credits were taken by Lynn Byczynski. Other photos are credited to their contributors.

About the editor

Lynn Byczynski is the editor and publisher of *Growing for Market* magazine, which she founded in 1992. Lynn and her husband, Dan Nagengast, were market growers for more than 20 years at Wild Onion Farm near Lawrence, Kansas. They grew vegetables for a CSA, farmers markets, grocery stores, and restaurants. They also produced cut flowers for florists, a pick-your-own subscription, and grocery stores. They erected their first hoophouses in early 2000 and soon had 8,000 square feet under cover. Lynn is the author of *Market Farming Success* and *The Flower Farmer: An Organic Grower's Guide to Raising and Selling Cut Flowers.* She is the co-author with Erin Benzakein of *Fresh from the Field Wedding Flowers.*

Lynn Byczynski and Dan Nagengast at Wild Onion Farm.

About *Growing for Market*

Growing for Market is a magazine for small farmers who direct market produce and flowers. In its 20+ years of publishing, GFM has provided a platform for growers to write about their experiences and ideas. Every article in GFM is written by a grower or other front-line practitioner. As a result, articles are always serious, practical, and credible. GFM covers every aspect of direct market farming — from production to marketing to business management. It is considered essential reading by sustainable market farmers throughout North America.

Growing for Market is available by subscription as a print or digital magazine. GFM also offers membership to its online archive of more than 1,500 articles published since 2001. In addition, GFM sells the best books about direct market farming; subscribers receive a 20% discount on retail prices. All books mentioned in this publication are available from GFM's book store.

Visit www.growingformarket.com for more information.

Preface to the second edition

The first edition of *The Hoophouse Handbook* was published in 2003 and revised in 2006. At that time, hoophouse production was still relatively new in the United States and hoophouse growers were feeling their way. Best practices were not yet established and there was a lot of experimentation happening on small farms all across the country.

At first, hoophouses were used primarily for season extension of field production in spring and fall. Eliot Coleman, owner of Four Season Farm in Maine, started experimenting with winter production in unheated and minimally heated hoophouses and eventually published the most important book on the topic, *Winter Harvest Handbook*. Across the U.S. and Canada, growers started extending the season later and later into winter. While winter production was taking hold, people also were realizing the potential for summer production. Hoophouses have been used for decades in Mediterranean countries for summer vegetable production and in Southern California for cut flowers.

As soon as farmers recognized the year-round potential of hoophouses, the floodgates opened. Growers in all climates built these inexpensive structures and started growing all kinds of crops in all seasons. Structures improved dramatically to fit the new uses. Protocols were established that more or less guarantee success. Even the U.S. Department of Agriculture got on the bandwagon and created a funding program to help farmers buy hoophouses. (Details on page 107.)

Today, there is no reason for a new hoophouse grower to fumble around in the dark, experimenting and sometimes failing. There is a clear route to follow. In this new edition of *The Hoophouse Handbook*, you will find that route to success explained and illuminated.

A few words about terminology

When the first edition of *The Hoophouse Handbook* was published in 2003, most growers called their structures hoophouses. Extension and university folks tended to call them high tunnels. Both terms refer to the same thing. Over the past decade, the term high tunnels has become predominant over hoophouses. But in the interest of preventing any confusion about the history of this book, we have retained the same title. We use *hoophouse* and *high tunnel* interchangeably throughout the book.

Whichever term we use, you can assume that we are talking about a greenhouse structure in which crops are grown in soil, in the ground. The structure may be any shape — quonset, Gothic arch, or gabled — and it may have one layer of poly covering or two layers inflated by a fan. It may or may not have electricity. Finally, although most hoophouse production is in unheated structures, this book will cover some production in heated structures. Again, the key distinction we make in *The Hoophouse Handbook* is that crops are grown in the ground, not in containers.

How this book is organized

In Part 1, you will read about crop production and the many ways growers are profiting from hoophouses. This section includes information on growing vegetables and cut flowers, in winter and the rest of the year. You will read what growers and researchers have discovered about the most profitable uses of valuable hoophouse space.

In Part 2, you'll read about buying and building a hoophouse. You will learn about the many types of structures, the lingo manufacturers use, optional features that you may or may not need. We'll also provide some tips on where to build and which direction to orient the hoophouse. Finally, we'll share some hard-won advice on laying out the structure and covering it with plastic.

In Part 3, you will learn about the many additional resources available to you. Since hoophouse growing has become so popular, there are virtually thousands of websites and publications that can help you get ideas for what will work best in your particular climate and location. We'll help you uncover the best of that information.

I hope that by the time you have read through this book, you will feel empowered to get started with hoophouse growing. With so many benefits awaiting you, there's no reason not to jump right in.

Lynn Byczynski
Growing for Market
November 2014

Part 1 Growing in hoophouses

Many growers have been amazed by the productivity of hoophouse growing. In one research project, the hoophouse crops were so much more profitable than field crops that the university researcher commented, "They ought to be called Economic Development Units." Another researcher, after just a few seasons of trialing crops in a hoophouse, asked the rhetorical question, "Why aren't we growing everything in these things?"

In fact, though most crops do well in hoophouses, there are some exceptions. Some do better in the field and some are only marginally better under cover, so they don't justify the expense of a hoophouse. In this article, I'll describe the benefits we have found on our own farm during 12 years of hoophouse production. In the articles that follow, you'll hear from growers across the country about what works for them.

At my farm near Lawrence, Kansas, we have grown in four hoophouses, each 20 x 96 feet. We are in Zone 6, with a frost-free season of April 20 to Oct. 15. Summers are extremely hot and winters can be extremely cold, though without a lot of snow. We have grown vegetables for a Community Supported Agriculture program and grocery store. We also have grown cut flowers for florists and grocery stores. Our goal for hoophouse production was to keep our markets satisfied over the longest possible season without using heat in the hoophouses.

To that end, our hoophouses do not have heat or electricity. They are covered with a single layer of 4-year greenhouse poly. In their first year, when we had erected just two, the value of the crops we produced was more than twice what we paid for the structures. We were so impressed that after nine months, we added the second pair.

Because we got started with hoophouse growing before it become commonplace (around the year 2000), there were not a lot of greenhouse structures specifically designed for growing vegetables and flowers in the ground. We chose to buy Stuppy's Polar Cubs, a relatively lightweight model normally used for overwintering container plants at nurseries. We went with Stuppy's because they are located nearby, resulting in lower shipping costs for the structures. Today, many vendors sell structures that are going to be shipped from a factory elsewhere so there may not be an advantage to using a nearby vendor. But ship-

At the author's farm near Lawrence, Kansas, four identical hoophouses are used to produce a wide array of vegetable and cut flower crops across all four seasons.

ping is a big chunk of the price, so be sure you know the shipping costs before you buy.

The Stuppy Polar Cubs have 40 inch sidewalls. We regretted that choice almost immediately because the low sidewalls limited the type of crops we were able to grow in the side beds. Lettuce and greens were fine, tomatoes and most cut flowers were too tall. We highly recommend getting a high tunnel with sidewalls at least 4 feet tall.

We made raised beds with a tiller, the same way we raise beds in the field. There are two beds, each 4 feet wide, down the center of the high tunnel, and a 30-inch-wide bed along each sidewall. That gave us comfortable 2-foot-wide paths between beds. Because we decided that hoophouse space was valuable, in one hoophouse we decided to reduce the footpaths and contain the raised beds with lumber. So we had two 5-foot-wide beds in the center and a 30-inch bed on each sidewall. We were definitely able to get more plants into the contained raised beds and we thought the expense of the lumber was worthwhile for extremely high-dollar crops such as certain cut flowers.

Temperatures

Early in our years of hoophouse production, a university researcher installed data loggers in one of our hoophouses so that she could track temperatures inside compared to outside at the same time. The data gathered was useful in proving that our experience was pretty much the same in winter as Eliot Coleman's experience in Maine, which he details in his book *Winter Harvest Handbook*.

In winter, when outside temperatures dropped to 10 below zero at our farm, the temperature inside our hoophouse was 4°F above zero. However, by the time it got that cold, we had already installed an inner tunnel of row cover on hoops directly above the cold-hardy crops. Beneath this second layer of protection, the soil never froze.

In spring, we found that temperatures fluctuated widely as the houses heat up rapidly on sunny days, regardless of the outside temperature. It was imperative that someone be home to roll up the sidewalls and open the doors on sunny mornings. Otherwise, the plants suffered heat stress such as bolting in vegetables and blooming on short stems in flowers.

Once the temperatures reach the 50s at night, we remove the endwalls and leave the sides rolled up. With enough ventilation, the inside temperature stayed the same as the outside temperature.

The author's hoophouse with larkspur and strawberries in early spring.

Chief benefits

Here is the list of benefits we observed from our hoophouses over the years:

1. They allowed us to get early tomatoes and to provide higher-quality tomatoes because of our ability to keep soil moisture constant; in the field, tomatoes are often subject to a host of physiological problems caused by uneven moisture.

2. They allowed us several extra weeks of tomato harvest in fall for two reasons: the plants remained healthy much longer than in the field and the high tunnels protected them against light, early frosts.

3. They allowed us to grow greens for restaurants until mid-December, and they kept the family table supplied virtually all winter long.

4. They allowed us to grow certain delicate flowers that can't handle wind, particularly the high-dollar flowers lisianthus, delphinium, and campanula.

5. They got us into the flower market in time for Mother's Day, when demand for flowers is at its highest point of the year.

6. Finally, and perhaps most important, the hoophouses give us a greater measure of control over our income. When nature sends a late freeze, early freeze, hailstorm, excessive rain, etc., we have a layer of protection on at least some of our crops.

Here is a look at the schedule of planting and harvesting flowers and vegetables in our hoophouses.

Fall-planted flowers

The first crops to go into the hoophouse in fall are flowers that will overwinter and bloom early in spring:

• Delphinium. I love the airy, clear blue 'Bellamosum' and the tall-but-manageable spires of 'Clear Springs'. I or-

Sweet William, a biennial type of Dianthus, is planted in fall for early spring bloom.

der both as small plugs in July and receive them in September, planting no later than Oct. 1. The plants grow well in fall and stay green under row cover all winter. We cut our first delphiniums April 26, and they produce through July. The 'Bellamosum' often rebloom in fall.

Delphiniums were one of the great joys of hoophouse growing for me. I had tried to grow them outside for years with no luck, so having superb-quality blooms for three solid months was gratifying.

• Sweet William. I start seeds myself in our propagation greenhouse in July or August or I purchase plugs for late September planting. There are numerous types of biennial dianthus, and I love them all. Favorites include the multi-colored mixes 'Tall Doubles' and 'Electron', and 'Newport Pink', which is really more of a coral color.

For fall planting, be sure to choose biennial types. There are also annual types such as 'Amazon' and 'Purple Bouquet' that don't require vernalization and should be planted in late winter/early spring in the high tunnel.

• Larkspur. We direct seed larkspur in the field in late September. We direct seed it in the hoophouse about two weeks later. The hoophouse larkspur is not as strong-stemmed as field-grown, but it's perfectly usable. And it's two weeks earlier than field-grown — often ready for Mother's Day sales. It blooms out quickly in the warmer hoophouse, lasting only two weeks, but by then the main crop is blooming outside.

• Campanula. The biennial Canterbury Bells should be fall planted because it requires vernalization to bloom in spring. The 'Champion' series is an annual that should be planted in early spring.

• Foxglove. Another biennial that needs vernalization to bloom in spring. Tall like delphinium, it needs plenty of room.

• Hardy annuals. There are a large number of flowers that direct seed themselves outside and therefore can be planted in fall in the hoophouse for early spring bloom. They are not very valuable flowers but they are great fillers for bouquet making. I like to plant small patches of Bupleurum, Nigella, Orlaya, and Saponaria.

• Tulips. Bulbs can be planted in late fall with no row cover; they need cold weather. They will bloom in early spring tall and strong. We just pull the bulbs as we pick, rather than trying to perennialize the tulips because they are never as nice the second year. It's better to use the valuable hoophouse space for spring and summer crops, then plant new bulbs in fall.

Fall-planted vegetables

By planting in mid September to early October, we have vegetables until at least mid-December, and the super-cold hardy arugula, mache, and spinach can be harvested all winter. Several crops can be planted in fall to get an extra-early start in spring. Here were the big winners:

• Spinach was direct seeded and covered with row cover over wire hoops. We started cutting in fall and went all winter into spring. Once the weather warmed up, aphids arrived and the spinach had to be plowed down.

• Cilantro was direct seeded and produced well in fall and again in spring. It's not a big seller, especially because people mostly like to use it with tomatoes, but it's very cold-hardy.

Hoophouse lettuce is crisp, clean, sweet, and colorful when grown in the cooler temperatures of spring and fall.

Lisianthus thrive in the heat of the hoophouse in summer.

• Arugula was direct seeded and covered with row cover. It produced quickly, about 30 days from seeding, then re-grew for a second crop. Arugula can be succession planted several times to ensure a steady supply all winter.

• Chinese cabbage got huge in the hoophouse. Although it seemed to take up too much space, when we ran the numbers on them even at the low wholesale price of $2 a head, they still trounced baby lettuce.

• Baby lettuce was pretty, with the red varieties especially attractive in the cold weather. But it grew very slowly and therefore didn't bring in a lot of revenue.

• Leeks were transplanted into the hoophouse in October and they overwintered nicely for an early spring harvest.

Spring-planted flowers

We have used the hoophouses for three kinds of spring-planted flowers: cool-loving annuals that need to be planted early in order to bloom before it gets hot; crops that can't tolerate wind; and late-bloomers that might get killed by early frost outside the following fall. My most profitable:

• Lisianthus. This most lovely of summer flowers is perfect when grown under cover. Outside, it achieves only 18-inch stems; inside, it's 36 inches tall. Flowers don't get spoiled by water spots and foliage stays disease-free. We routinely get a second flush of blooms after the first flush is harvested. We have even left the plants over winter with very little loss, and had them bloom again the second year.

We bought plugs and planted them in the hoophouse between March 25 and April 12. Whatever the planting date, they started blooming June 25.

• Campanula 'Champion'. Like lisianthus, it never got very tall outside but was tall and strong in the hoophouse. We bought plugs and planted them in late March. They tend to bloom throughout June then die out from our heat.

• Sunflowers. We grow the single-stemmed hybrids, particularly the day-neutral varieties 'Sunrich', 'Pro-Cut' and 'Sunbright Supreme'. We start seeds in the heated greenhouse March 5 and transplant to the hoophouse March 25 for bloom in early May. We do several successions in the hoophouse to get us through Mother's Day. But as day length increases, sunflower plants get bigger — so we grow these space hogs outside in summer.

• Bells of Ireland. We planted them from plugs one fall and got a nice spring crop. But we obviously let some go to seed because now, many years later, we have bells popping up everywhere in the high tunnel as soon as the temperatures go down into the 40s. They love cool weather.

Spring-planted vegetables

The main hoophouse crops for most vegetable growers are tomatoes and cucumbers, and our experience has validated the benefits of growing these two under cover. Our tomatoes ripen one to three weeks earlier and the cucumbers two weeks earlier than field-grown crops. Yield and quality are much better in the hoophouse. Many heirloom tomato varieties that struggle outside do much better inside. Foliage stays healthy, so the same plants will produce all summer and into fall.

Tomatoes and cucumbers both require trellising and they both take up a lot of space. Even so, revenue per square foot is as good as cut flowers and better than every other vegetable we have grown. And tomatoes are one of those crops that can help you sell everything else — if you are the first person at market with ripe tomatoes, customers will flock to you.

Peppers also may benefit from the hoophouse, not so much because of earliness as because of protection from sunscald. The price, however, may not justify the space.

Summer-planted flowers

After the spring flowers have capitulated to the heat, we have space in the hoophouse for late plantings of crops that we can protect from frost in autumn. We choose heat-loving varieties because it can be sweltering in there in July and August, when we set out the transplants. The new plants require a lot of water to get established.

• Salvia leucantha. Also known as Mexican sage, the fuzzy purple spikes of flowers are the perfect compliment to autumnal colors of mums and sunflowers. Salvia leucantha requires short days to bloom, so it's not ready for harvest until late September. It is extremely frost sensitive, though, so even a light frost in early October can kill it. In the hoophouse, it usually survives the first few frosts of fall and we get a month to six weeks of production from the huge plants.

• Celosia. Another heat-lover that will bloom long into fall if protected from light frosts. We plant several types, both plumosa and spicata varieties, and lean toward autumnal colors of orange and red for our fall bouquets.

Most profitable uses for that valuable space

By Lynn Byczynski

When USDA's Natural Resources Conservation Service announced in 2009 that it would provide cost-share money for growers to buy one hoophouse (See page 107), it raised an interesting question: If you had only 2,000 to 3,000 square feet of hoophouse space, what would be the most profitable way to use it?

Early hoophouse users, including me, found that sales of hoophouse crops covered the cost of the structure in one season. But the price of a hoophouse has increased significantly in the past decade or so, and it is no longer a given that a hoophouse will pay for itself immediately.

In fact, David Conner, an agricultural economist at Michigan State University, had a grant to give nine farmers hoophouses and then study how much revenue they were able to generate in them in a year. Sales ranged from less than $1,000 to $8,000. Net income ranged from -$500 to $5,000. Because the hoophouse structure cost $10,000, the payback would be two years for the most successful grower in the study.

In order to increase your chance of profitability and get that hoophouse paid off, it is helpful to consider the experiences of other growers and researchers. In a nutshell, it looks like tomatoes, winter greens, and cut flowers are the top hoophouse crops. But, as with everything agricultural, actual results may vary. You should consider the context of the research and compare it to your own climate and markets, then judge whether a crop will be profitable for you.

Here is a roundup of some research projects on hoophouse profitability:

High tunnel blueberries put on an autumnal show of color at the Kansas State University research station near Kansas City.

Nationally, it's tomatoes

Ted Carey, formerly a vegetable specialist at Kansas State University, was the lead author of a report about high tunnel use nationwide, published in 2009 in *HortTechnology*. The researchers sent a survey to horticulture/food crop Extension specialists in every state and received responses from 46 states. They asked about the acreage of high tunnel production and most popular crops.

For most states, tomatoes were the #1 crop produced in high tunnels. Winter greens were important in many states, as were cut flowers. Small fruits also ranked high. "High tunnels are amenable to large-scale production, with raspberry production in California being the most striking example, but in most states high tunnels are being used by relatively small-scale producers to serve local markets," the report says.

California was by far the largest high tunnel production state, with more than 4,000 acres of multi-span tunnels growing primarily raspberries. Oregon came in second with 100 acres of mostly berries, and Florida reported 50-100 acres of primarily Oriental vegetables, leafy greens, and berries.

On the subject of tomatoes, Ralph Cramer, a Haygrove High Tunnel dealer, said that one grower reported getting 68.8 tons per acre of grafted determinates. That is more than double the estimated yield of 25 tons per acre for field-grown tomatoes.

Carey said that many small growers use the tunnels to produce both early tomatoes and winter greens. In most places, that means the tomatoes have to be removed in September to allow time for greens to mature before the short days of winter arrive. That can be tough to do, when tomatoes are still bearing well. But it is possible to let the tomatoes produce until first frost, and still get a spinach crop planted. Carey and Sharon Knewtson, a graduate student, studied spinach yield from various planting dates at their research station near Kansas City. They found that spinach planted in October can be harvested as soup-spoon sized leaves throughout the winter without significant loss of spring yield. Planting in mid-November, however, led to reduced spring yields.

Tomatoes in Iowa

Iowa State University researchers collected financial data from five separate hoophouse growers and created crop enterprise budgets to calculate returns per square foot of the various crops the farmers produced one year between April and October. The researchers, Linda Naeve and Craig Chase, adjusted the data from the different farms to make them comparable; for example, they averaged the sales price when more than one farm grew a crop. Their numbers assumed the entire crop was sold, but they did not include marketing costs in the profit calculations. On the expense side, they calculated inputs, labor, and ownership costs — the cost of the high tunnel divided by its eight-year life expectancy and the plastic by its four-year life. To understand all the parameters used in this study, it's helpful to read the complete report at https://store.extension.iastate.edu/Product/Vegetable-Production-Budgets-for-a-High-Tunnel

"Overall, the potential annual returns for a high tunnel (excluding marketing costs) based on the findings of the five farms would indicate that a multi-crop or tomato high tunnel enterprise could be approximately $3.00 per square foot or $6,480 for a 2,160 square foot tunnel (30 ft. x 72 ft.). Assuming an initial investment of around $7,000, this level of return would indicate about a 1-year payback, which is unusual in agriculture."

The tomato high tunnel numbers were based on a price of $1.60 per pound for slicing tomatoes. The multi-crop tunnel included slicers and grape tomatoes, bell peppers, lettuce, herbs, greens, eggplant, and cucumber. The highest gross sales per square foot were herbs ($25.60 per square foot from just 9 square feet) and grape tomatoes ($10.80 per square foot from 127 square feet).

Cucumbers were the least profitable, netting just 38 cents per square foot. Bell peppers returned 88 cents per square foot.

When looking at raw numbers, it's important to remember that you have to be able to sell everything you grow to get comparable returns. A high tunnel full of herbs has the potential to produce more than $55,000 in sales — but only if you have a market for that volume of herbs.

Greens in Michigan

Michigan State University has some of the most extensive experience in high tunnel growing for market on its Student Organic Farm. The SOF has 4 acres of production, including numerous unheated high tunnels. It sells through a 48-week CSA, a six-month farm stand, and to the university's dining service. John Biernbaum, professor of horticulture, has been keeping data on the amount and value of food the farm produces. Data from 2008 reflect production in 11,500 square feet of high tunnel. But only about 7,000 square feet of that space is cropped because of the need to have wide paths that can accommodate classes and tours.

In 2008, the farm had $120,000 in sales from the 4 acres. Of that, about $40,000 came from the high tunnels. Although at least 24 crops were grown in high tunnels, the majority of the revenue came from four: tomatoes, salad greens, lettuce, and spinach.

Biernbaum cautions that the dollar amounts are somewhat soft because most of the produce goes through the CSA, and because there really are no comparables for the winter greens. Also, he hasn't calculated revenue per square foot on the crops yet. But at the $10 a pound the university pays for salad mix, he thinks that's likely to be the top-grossing crop. Tomatoes also appear to be highly profitable per square foot.

"I think the data will show it's possible to make $5 a square foot gross sales," he said.

In Michigan, winter crops are planted into the hoophouse in August and September for winter harvest.

"Careful planning is necessary to have the houses full going into late November," Biernbaum said. "At the SOF, the majority of the bed space is ready to harvest in November. That has to get us through CSA distribution until Dec. 15, and then starting again Jan. 15 until growth picks up in February. That means as many as 8 to 10 distributions worth of greens and fresh harvested crops to go along with the potatoes, squash, cabbage, carrots, garlic and other items from storage. Our current planting plan and rotation has been in place for three years."

"Well managed tomatoes can clearly provide significant income, but winter harvested salad and cooking greens and fresh crops are important for local food and healthy people," Biernbaum said.

Small fruits

Blueberries, raspberries, blackberries, and strawberries are also popular crops for high tunnel production, but for different reasons in different climates.

Curt Rom at the University of Arkansas has been researching organic berry production in high tunnels for three years. Here's what he says about the benefits in his state:

"For berries, there is the potential to advance the harvest season of early ripening cultivars by 2-4 weeks. Likewise, we have seen the ability to extend the fall season by several weeks for fall bearing raspberries and blackberries. The greatest advantage we have seen with raspberries. Raspberries are not adapted to the hot south and fruit will not mature. With the tunnels, slightly reduced light levels in the tunnels and with fruiting earlier or later in the season, we have avoided some of the excessive heat.

With both early and late season berries, it become imperative to have frost protection within the tunnels. We have found a combination of frost curtains (poly-fabrics over the crop plants) and supplemental heating allowed us to survive temperatures 10 degrees below freezing.

"The greatest advantage we have seen for the berries is keeping the rain off the crop. This spring we had almost crop failure in the field grown berry plants due to poor pollination (too cool and wet during the bloom and fruit set period) and with heavy spring and early summer rains, we had poor fruit quality in outside plots. However, we avoided all of that in the tunnels. The advantage is you never lose a day to rain - fruit set, crop work, or harvest. And, fruit are not ruined by the rain."

In the Northeast, the same benefits are reported. In addition, researchers have found that yield of fall-fruiting raspberries is double that of field-grown raspberries in part due to a lower incidence of gray mold disease. Cornell and Penn State have published a 35-page guide, *High Tunnel Raspberries and Blackberries*, with recommendations for varieties and scheduling.

Fall raspberries

Fall-bearing raspberries are one of the success stories of hoophouse research under way at Penn State University. Researchers knew that early frosts had reduced yield of fall raspberries by as much as 85%, so they planted raspberries in hoophouses with the goal of extending the fall picking season. They grew 'Heritage' and 'Autumn Britton' and planted them in the hoophouse about a month before they could have planted outside. Harvest lasted three weeks longer in the hoophouse than outside, and yield on the first-year plants was nearly as high as that which could be expected from a three-year-old field planting. Ninety-eight percent of the fruit harvested was marketable. No fungicides were used in the hoophouse, and shelf life was more than double that of raspberries grown in the field under a standard fungicide spray program. The biggest problem was spider mites, which were controlled with predatory mites.

Garlic

Twelve garlic varieties were grown in a hoophouse at Penn State University's Center for Plasticulture. All were planted Oct. 25-31 and harvested in late June, except 'Xian', which matured 2.5 weeks earlier. Bulbs were larger in the tunnel than the field, but no other comparisons were made to field-grown garlic. The researchers found that high tunnel garlic returned 10 pounds for every 1 pound planted and was generally of better quality.

In northern Minnesota, however, researchers showed that high tunnel garlic was much more successful than field-grown garlic during very cold, dry winters. For three years, survival rates of hardneck garlic 'Music' ranged from 2% to 79% when grown outdoors. Inside a high tunnel, survival rates ranged from 87% to 95%. The year that field-grown garlic had only a 2% survival rate, there was little snow cover and the temperature went as low as -41°F. Inside a Gothic arch high tunnel that winter, 92% survived and in a round high tunnel, 87% survived.

The researchers also found that softneck varieties, which are normally not grown in very cold areas, had at least an 85% survival rate in a high tunnel.

Satisfied growers in OK, TX

Steve Upson at the Noble Foundation in Ardmore, Oklahoma, surveyed hoophouse owners in Oklahoma and Texas about their crops, structures, and revenue. Eighteen growers with 34 hoophouses participated in the survey.

"In the business world, the percentage of repeat customers is a direct indication of product satisfaction," Upson wrote in his report on the survey. "How likely are growers who have hoophouses to construct a new one? The survey said 89 percent are very likely or somewhat likely while only 9 percent said "no way."

"Some of the reported benefits of hoophouse growing over field production include increased yield, improved crop quality, risk management, profitability, competitive advantage, convenience and season extension. When asked how satisfied they are with these benefits, over 70 percent said they were very satisfied with the increased yield, improved crop quality, earlier production and competitive advantage. Only 56 percent of the respondents said they were very satisfied with the level of profitability associated with hoophouse production while 31 percent were somewhat satisfied. Growers should not be fooled into thinking hoophouse production is a get-rich-quick scheme. Yield increases and earlier production come at a cost — the cost to construct and maintain the house."

Patriotic potatoes

A 17 x 96 foot unheated hoophouse can produce up to $1,278 in revenue in less than three months when planted with specialty red, white and blue potatoes. William Lamont Jr. of Penn State University reports that he planted the patriotic-themed spuds on April 6 and harvested them on June 27. The potatoes were planted in a typical plasticulture system, including plastic mulch, drip tape and row cover, inside the hoophouse. The potatoes were planted in four double rows, 13 inches apart, spaced 8 inches in the row. No pesticides were used and the quality of the potatoes was excellent.

Yields (extrapolated to represent one hoophouse full of each variety) were 1,104 pounds of Red Pearl, 920 pounds of Eva, and 1,278 pounds of Michigan Purple. Lamont estimated that growers should be able to direct market the potatoes for the Fourth of July.

Why winter growing works

By Eliot Coleman

Sunshine, and the warmth it transmits, are two key ingredients for plant growth. Wouldn't it be wonderful if in the heart of food-producing middle America, say, some place like Lawrence, Kansas, we had as much winter sunshine as sun-drenched Mediterranean paradises like Majorca or Corfu? Then there might be a chance for commercial growers other than the Californians to easily harvest a wide range of crops right through the winter months to supply their local markets. What a shame most of us live in the far north where short winter days limit the sunshine.

But hold on a minute. Lawrence, Kansas, despite occasional snow and cold weather, is not in the bleak far north. The hometown of *Growing for Market* lies on the 39th parallel of latitude. All places on the same parallel of latitude have the same day length. The 39th parallel runs across the United States from Sacramento, California, out in winter food production country, through Lawrence, to Washington, D.C. On the other side of the Atlantic, believe it or not, it runs through Majorca and Corfu. If only we had their moderate climate. Well, there's an easy way to deal with that obstacle.

The winter greenhouse vegetable industry appreciates good sunshine and is very active along the 39th parallel. But you don't need high-tech systems or fossil fuels to harvest winter vegetables. There is an unconventional winter garden alternative for commercial growers that will transport your plants to a warm and sheltered land. It is both simple and resource efficient. Any greenhouse, from simple hoop structures to fancy glass conservatories, which grows plants in the soil can produce fresh winter salads. No artificial heat is necessary. No complicated heat storage system is necessary. All you need is to add a second protective layer of translucent material inside the greenhouse. One of the spun-bonded garden fabrics held a foot above the soil on wire wickets is perfect. Under this protection-within-protection, what I am calling "twice-tempered," the inner climate is moved three USDA zones to the south and hardy winter vegetables like red oak-leaf lettuce, spinach, arugula, mache, 'Bull's Blood' beet, claytonia, minutina, parsley, chard, scallions, tatsoi, pak choi, turnips, carrots, and many more will thrive. Yes, they still freeze on cold nights but these hardy crops are not damaged by below-freezing temperatures as long as they are safe from the killing wind-chill, alternate wetting and drying, and ice storms of the outdoor winter. When they thaw during the day they look like they're growing in a perpetual spring.

This idea, protecting hardy winter crops with a second layer inside an unheated plastic greenhouse, was pioneered in the 1950s by Professor of Horticulture E.M. Emmert at the University of Kentucky. For some reason, no one picked up on Emmert's innovation at the time. Possibly because plastic greenhouses were so new; possibly because the concept seemed too good to believe. But more likely, growers were deterred by unfamiliarity with the importance of the day-length factor in winter gardening.

At Eliot Coleman's Four Season Farm on the coast of Maine, stationary and rolling high tunnels plus low tunnels make winter growing feasible.

Plant in fall

Because of the slowdown in plant growth due to cooler temperatures in shortening fall days, winter crops have to be planted before winter. Timing the plantings is the key to success. In the Lawrence, Kansas, area, for example, the seeds for most of the hardy salad crops (oak-leaf lettuce, spinach, mache, arugula, claytonia, etc.) should be sown from mid-September to mid-October. Other crops like carrots, parsley, scallions, and chard should be sown about a month earlier. The goal is to get the plants alrge enough so you can cut mesclun-sized salad leaves or pull baby carrots before the day-length becomes shorter than 10 hours. According to the Old Farmer's Almanac, the period when days are shorter than 10 hours at the 39th parallel of latitude extends from November 16 to January 24. During that time, the growth of newly sown seeds is extremely slow. But the crops that were established earlier play be different rules. They sit there happily in the protected conditions waiting to be plucked for market. Those crops that can be cut more than once (spinach, lettuce, claytonia, 'Bull's Blood' beet, etc.) will produce new leaves at a slow but consistent rate during the short-day period, then magically begin a new burst of growth in response to lengthening days starting the last week of January. You can sow new seeds in empty spots any time you want. These will grow slowly at first and then come on quickly to continue your harvest through the spring.

Plant in January

In fact, that's a good way to start right now to give this idea a trial. If you have a plastic greenhouse (hoophouse, high tunnel) that you use to get on the market with early crops, put in an inner layer, start planting in January, and plan to be off and running sooner than ever. In our Zone 5 climate in Maine, we sow our first carrots on January 1 with three more succession plantings at two-week intervals. We use the pinpoint seeder and plan on 6 to 8 square inches per carrot for the first seeding and 4 square inches thereafter. By May 10 we are selling bunched baby carrot varieties like 'Primo' or 'Mokum' to a very appreciative market, a full seven weeks ahead of the best we could hope to do outdoors even with row covers. We begin our pre-sprouting of seed potatoes on Feb. 15, plant them under the inner cover on March 15 (two rows to a 30-inch bed) and are selling baby new potatoes along with the early carrots on May 10, again to raves from our customers who have been hungering for fresh spring crops. We have transplanted broccoli under the inner layer in March (again two rows to the bed) and sold May brocoli.

We keep producing our mesclun salad mix all winter with this system but even if you just start planting your mix in January you should be back on the market by late March. This will be some of the highest quality mesclun you have ever grown because these salad crops prefer the cooler temperatures. Remember to keep the day temperatures cool by venting both the greenhouse and the inner covers on sunny winter days.

So give it a try. Consider September a second spring for planting winter-harvested crops, and January as the start of a new season. This doesn't just apply to Lawrence, Kansas. Most of the U.S. has as much or more winter sun than the south of France. (Portland, Maine, for example, is on the same parallel of latitude as St. Tropez.) The twice-tempered climate in your greenhouse is three zones warmer than where you live. Use it to expand your fresh marketing season into months you only dreamed of.

Eliot Coleman has more than 40 years of experience in all aspects of organic farming, including field vegetables, greenhouse vegetables, rotational grazing of cattle and sheep, and range poultry. He is the author of The New Organic Grower, Four Season Harvest *and* Winter Harvest Handbook. *With his wife, Barbara Damrosch, he operates Four Season Farm, an experimental market garden in Harborside, Maine. The farm produces vegetables year-round and has become a nationally recognized model of small-scale sustainable agriculture. Learn more at www.fourseasonfarm.com.*

Eight tips for winter success

By Ben Hartman

At the cutting edge of the local food movement is the challenge of providing fresh, local food even in the middle of winter. Growing in the winter costs more (greenhouses/high tunnels and row covers aren't cheap), but when best practices are employed, might it still be possible for local growers to compete with out-of-region food producers?

In August 2014 in Fairlee, VT, 25 growers with experience in winter production, along with 10 others representing seed companies, Extension programs, and universities, met to discuss these best practices. The group spent two days unraveling what in the past several winters has worked well, what hasn't, and where we all need to spend more effort researching and experimenting.

The idea to pull together such a group came from Eliot Coleman, farmer, author, and long-time advocate of winter growing, who along with daughter Clara Coleman and Sandy Arnold (Pleasant Valley Farm, Argyle, NY) organized the gathering, called Frozen Ground.

When aspiring farmers visit our place, I am often asked, "Where's the best place to learn how to farm for a living?" My answer: "There is no substitute for learning directly from other farmers making a living from their food." So I was eager to hear from this group of winter farmers, as well share our own experience. Winter growing is a significant and growing part of our farm sales (in the January 2014 issue of GFM I wrote about how our crops fared this past winter).

So, what worked? In general, growers found greenhouse-grown spinach, kale, arugula, swiss chard, and Asian greens (baby and full size) to be consistent winter performers. Of these, spinach was the favorite, and some farms in the winter grow spinach exclusively. Baby lettuce and head lettuce is less hardy, but with good management (see below), possible to grow. Some growers took fresh greenhouse radishes, turnips, carrots, leeks and other hardy crops to winter markets, while others said they preferred growing these in the field in the Fall and cold storing them for winter sales. These growers saved the greenhouses for higher-dollar-value greens. In general, kales and lettuces for full-size production were seeded in August and transplanted in September, while crops for baby production were seeded successively for 2-4 weeks starting mid-September. Johnny's Selected Seeds has developed a handy chart to help you figure out ideal seeding times for your area, available at johnnyseeds.com.

Warning: the seeding dates above are general guides (most growers were in zones 4-5). Don't take them to heart! The dates that will work best for your farm will be specific to your exact zone/microclimate and conditions (for instance, whether you have no heat, ground heat, or air heat). And keep in mind that winter weather varies considerably from year to year. The best practice to increase your chance of success is to sow crops over a 2-3 week period, then take notes! One of the most-used charts on our farm is what I call our "seed-to-harvest" chart—a record of dates we seeded and then harvested our winter crops. Now, if I want to plan for a spinach harvest in mid-March, I know when to seed. (The Powerpoint link at the end of the article includes our winter greens seed-to-harvest chart.)

Just as important as crop selection and seeding dates, however, is proper management during harsh winter weather.

Farmers are a creative lot, and every farmer at the gathering seemed to have a unique management approach to the common problems of cold temperature and low light levels. Even so, there was some common ground. For growers wanting to improve their success rate in the winter, or those wanting to start tinkering with winter production, below are eight management tips that experienced growers use to guide their winter production.

1. Ventilate, ventilate, ventilate.

A challenge with greenhouse growing any time of year is high moisture, which leads to condensation on leaves and high incidence of foliar disease. The problem is exacerbated in the winter because of shorter days and lower light levels.

How does condensation work? According to Vern Grubinger, vegetable and berry specialist at the University of Vermont, "Warm air holds more moisture than cold air." In the winter greenhouse, Grubinger suspects that "condensation may occur in the evening as air cools and moisture drops out." Since nights are long, that condensate sits around for quite a while, leaving plenty of time for foliar diseases to develop.

The remedy is air exchange—pushing wetter air out and pulling drier air in. Some growers said they prefer "leaky" greenhouse rather than tight greenhouses, others keep a top shutter open all winter. We turn our endwall fans on regularly to exchange air.

At Screamin' Ridge Farm in Montpelier, Vermont, row cover is suspended on hoops directly above the plants. Vern Grubinger of the University of Vermont says that most growers find that running beds lengthwise is most efficient. Photo courtesy of Vern Grubinger.

Anytime a greenhouse leaks, you lose heat, but far worse to lose an entire crop to mold for lack of fresh air.

2. Keep inner covers as close as possible to plants.

Growers differed a lot in how they used inner covers and what type of covers they used, though after a lengthy discussion it was generally agreed that the closer a cover is to the crops, the better. The principle is simple: an inner cover is essentially a thin layer of insulation trapping heat released from the soil close to your plants. The less air space to keep warm, the better. One grower put it this way: "Would you be warmer with a blanket directly on top of you or suspended a foot over your body?"

Most, if not all, growers used some type of wire supports to keep row covers suspended above fragile new transplants like kale, chard, and lettuce. But beyond that practices varied widely. A common approach was to float covers one foot to 18" above crops to avoid damage from covers freezing onto plants. But others float covers directly onto their greens even in freezing temperatures, taking care not to remove them until plants have thawed.

The weight of inner covers varied among growers too, though most preferred a mid-weight cover (like Agribon 30) because that weight is strong enough to handle without ripping while still allowing some light through. A good practice is to add layers as the weather gets colder. At Pleasant Valley Farm, the Arnolds use one sheet when temperatures are in the 30s, two in the 20s, and 3 below 20°F. Once again, though, what works varies. It is worth experimenting on your farm with the height of your covers, the weight, and the number of covers needed to protect your specific crops.

3. Uncover inner covers on sunny days.

There are several reasons growers said it is worth the effort to remove row covers on sunny days:

Plants receive a maximum amount of sunlight, encouraging darker leaves and healthier growth.

The soil gets direct UV exposure, causing it to warm. When the covers are replaced at night, that warmth is released into the air between the ground and cover to keep the crop warmer

Improved air flow. More air movement means less condensation, which means less disease. Horizontal Air Flow (HAF) fans help reduce condensation and keep row covers drier.

Less risk of freeze up. Condensed moisture left on plants going into below-freezing nights can turn to ice, which will kill some plants and damage others. It's better to dry off plants as much as possible going into arctic nights by uncovering row covers during the day, thus exposing leaves to drying air. For the same reason, it is important to keep row covers dry by removing them when irrigating.

At a minimum, covers should come off two to three times per week, even if there is no sun.

4. Harden off plants in the fall.

Your plants will survive the depths of winter best if they are exposed to colder and colder temperatures as winter approaches. Some growers leave end walls open on cold nights, others keep their entire greenhouse (even sidewalls) open until the last possible moment (when the air becomes so cold as to damage crops). Depending on the crop, this was generally around the mid- to high-twenties.

Tom Stearns, owner of High Mowing Seeds, reported on test plots in a greenhouse in which staff at High Mowing tried to kill hardened-off lettuce via cold exposure. The temperature in the greenhouse was allowed to dip to 20 below zero, and they found several varieties that survived, even without row cover protection (the greenhouses were closed so there was no wind). Spinach was impossible to kill.

5. With supplemental heat, average temperature is what counts.

Growers using supplemental heat are looking to boost re-growth rates in their plants as well as protect plants from extreme cold. Growers reported using a wide range of low temperature settings for air temperature—from the mid-twenties (for hardy greens) to 64 degrees (for microgreens)—in order to encourage speedier growth. It was generally agreed that 28 degrees was a "safe" low for most cold hardy varieties.

However, it is the average temperature in a 24 hour period that will determine plant growth, not the high or the low. This principle can be applied to save on heating costs.

If you are looking to boost re-growth, it might pay to turn heaters on during the day--for example, by heating a greenhouse from 50 degrees to 65 degrees. This raises the average 24-hour temperature more economically than turning heaters up a similar amount at night when temps are colder and air is more expensive to manipulate (though you still want to keep plants alive at night!). Some growers use supplemental daytime heat to warm the ground enough to prevent the ground from freezing at night.

In general, growers using supplemental heat found that

At Pleasant Valley Farm in Argyle, New York, Paul Arnold and his daughter Kim harvest greens in a cold high tunnel. Photo courtesy of Sandy Arnold.

plant re-growth was indeed faster (in our case, 2-4 times faster) than in unheated greenhouses, but since heating costs vary widely (we have relatively affordable natural gas), there was no consensus on whether or not extra heat pays off.

Another practice growing in popularity is to bury hot water lines (black poly lines are sufficient) 18" below the soil surface and run hot water (around 110 degrees) through the lines as a way to heat the soil. Heaters ranged from geothermal systems to boilers (gas, wood pellet, corn, etc.) to tankless hot water heaters to hot compost heaps (in which black poly lines are coiled in the middle of a compost pile that sits just outside of the greenhouse—heated water is circulated with a pump). Results with these systems were mixed, with some growers reporting success, and others warning that heated ground equals moist air releasing from the ground, causing condensation problems I talked about in point #1. Some growers using ground heat said that for winter greens they prefer to keep the ground heat off during the darkest months of the winter when condensation is especially problematic. Plus, heat alone does not make plants grow, you need light, too. So boosting heat too much when days are short and cloudy is a waste of energy.

6. Use plastic mulch.

In the same way that plastic mulches warm field soils, increase earliness and increase yield, those same plastics heat winter greenhouse soils, increasing growth in the winter. The added benefit for winter growers is that the plastic traps mildew-inducing moisture in the soil (remember, when soil warms, it releases moisture in the air).

Of course, plastic won't work with closely-seeded baby greens, but for head lettuce, kale, and other transplanted crops, plastic increases earliness and yield. Some growers even transplant spinach into plastic, using plugs that contain 3 seeds per cell.

If you are growing summer crops in plastic in your greenhouses, you can reuse the same plastic for winter crops! Just pull your summer crops, poke new holes, and plant.

The two types of plastic being used in this way are the IRT (infrared transmitting) plastics and black plastic (biodegradable and non-biodegradable). IRT plastic is pigmented (usually green or brown) to allow some UV transmittance, which does a better job of heating the ground while still minimizing weed growth.

7. Plan ahead for chickweed.

An audible groan must have been heard across Vermont when the subject of chickweed came up. The sprawling weed is the nemesis of winter growers because it loves the cool, low-light conditions of the winter greenhouse. Though there was no Holy Grail solution for chickweed control, growers offered several techniques to try:

Flame-weed: Prepare your beds after summer crops, then wait two weeks, which allows chickweed to germinate, then seed directly over the chickweed. Before your winter crop germinates, torch with a flame weeder. Multi-torch flame weeders are available at flame-weeders.com.

"Cook" chickweed by covering ground with clear plastic in late summer.

"Freeze" chickweed by leaving greenhouses open during the coldest part of the year.

Use plastic mulch (see #6 above).

Use wire weeders and weed early, when the winter crop is only 1-2" tall and the weeds are in white thread stage.

Prevent seeds blowing in from outside the greenhouse. On our farm, we mow around all of our greenhouses for this reason.

Test compost for weed growth before spreading it in the greenhouse.

8. Manage fertility but keep nitrate and salt levels in check.

Growers using their greenhouses for summer production often want to boost soil nutrient levels to maximize yields of summer crops. According to Grubinger, "While winter crops still require an adequate supply of essential elements, they do not require the high nutrient levels needed to support fast-growing, large biomass crops of summer." Those high nutrient levels—especially excess nitrogen—can be a problem for winter crops.

Most field soil tests will not reveal nitrate levels, since they change rapidly depending on environmental conditions. Nitrates are inorganic ions that occur naturally as part of the nitrogen cycle. When microbes break down animal wastes, ammonia is created, which then oxidizes into nitrates. Nitrates are required for plant growth, but excess nitrates can stunt root growth, and plants that take up too much nitrogen invite pests like aphids.

Grubinger's advice is to test your greenhouse soil for nitrate-N by requesting a separate soilless media test, often called the saturated media test, which is designed for potting soils. This test is available from most land grant soil test labs, and it also measures the level of soluble salts.

The simplest solution to managing nitrate and salt levels is prevention—take soil tests regularly, and adjust fertility practices accordingly, which usually means dialing back your usual rate of fertilizers and other amendments. On our farm, we discontinue manure-based composts in the winter greenhouses. Other growers "flush" excess nitrogen and salts by irrigating heavily before transplanting winter crops.

In the greenhouse salts can become a problem because there's no rain and snow to leach them out of the soil. Salts include many of the nutrients that are contained in compost, manure, and fertilizers, and salts can be found in irrigation water. Over time, these can build up to levels that can injure plants. Salts are most pronounced at the surface level as water evaporates and leaves them behind, so they show up as white specks on the soil surface.

Grubinger suggests that growers test their irrigation water to determine its salt content. "When you change the plastic," says Grubinger, "leave it off for a period of time to allow salts to be flushed away by rain." It requires a lot of water to leach excess salts through flushing and drainage. If salts are not extremely high, another solution is to dilute them with the addition of new soil or peat moss. An extreme measure would be to replace the top couple inches of soil.

There wasn't time in the two days to cover everything related to winter food production. I yearned for a lengthier discussion of whether taking extreme measures in the winter (especially using expensive alternative heat systems) was truly paying off. I talked to one farmer who told me that while winter greens didn't lose money, they didn't make a lot either: he used them mainly to draw in customers to purchase more profitable storage crops. For others, however, sales of winter crops rival sales in the summer, with some winter greens in unheated tunnels giving returns of $8-9 gross per square foot.

Competition for summer market growers is getting tougher and tougher every year. Many areas are saturated with growers and at the same time there are more new growers coming on every year. In areas where summer production is crowded, winter growing offers a promising new market.

While winter growing is relatively new, it is out of its infancy, and meetings like Frozen Ground should help motivate seasoned growers to keep working out winter production kinks while giving new growers solid information as they get started.

For additional resources, many of the Powerpoints from this conference are posted at http://www.uvm.edu/vtvegandberry/WinterGrowingConference2014.html. An excellent guide to get started in winter growing was written by Becky Maden (Intervale Community Farm, Burlington, VT) for High Mowing Seeds: http://www.highmowingseeds.com/blog/growing-and-storing-for-winter-markets/. Vern Grubinger can be reached at vernon.grubinger@uvm.edu. Sandy and Paul Arnold can be reached at Arnold.pvf@gmail.com. Many of the farmers who attended the conference would be glad to give more information; see the Powerpoints for their contact information.

Ben Hartman is the co-owner of Clay Bottom Farm in Goshen, Indiana. Contact him at claybottomfarm@gmail.com.

Hoophouse crops for winter harvest

By Pam Dawling

The body of knowledge about using hoophouses for cool-weather crops has grown each year, and the options are many. Each grower will want to customize the mixture of crops to suit the local climate, market, and desired work schedule. For example, our focus is on supplying a wide range of fresh produce all winter and spring to feed 100 people. If your goal is to maximize harvests in December, and then read novels by the woodstove till mid-February, you might rather plant a large amount of spinach to overwinter and harvest after your time off.

The author's Zone 7 hoophouse produces all winter.

At our farm in central Virginia (Zone 7), we grow salad crops, cooking greens, turnips, radishes and scallions in the hoophouse. We also use some of the space to grow bare-root transplants for setting outdoors in February and March. Our outdoor beds can provide spinach, kale, collards, and leeks all through the winter, although the rate of growth doesn't compare to what happens in the hoophouse. We aim to be able to harvest greens in the hoophouse after the outdoor crops slow down, and turnips after the stored outdoor fall turnips have all been eaten, or as an occasional delectable alternative.

We have a 30' x 96' Clearspan house from FarmTek, the gothic arch type, with bows 4' apart. We use an inflated double layer of plastic. Our hoophouse does not have roll-up or roll-down sides. We built it for maximum winter coziness, and so only the ends open. We don't use row cover over our winter crops. We like that the insulation provided by the double plastic means we can move around freely and work inside the tunnel, unhindered by row covers. We don't need them for warmth, and in winter the amount of daylight is a limiting factor, and row cover does reduce the light. If you are in a colder climate than we are, the extra benefit of row covers can pay off. Compared to using row cover outdoors, inside the hoophouse is easy

– no gusts of wind, no rain, no snow!

Inside, we have seven lengthwise beds. The two edge beds are 2' wide and the rest 4'. The paths are 12". This is the same size we use outdoors, and works well for us. It really isn't much path space. (Do yoga! Or make wider paths.)

We do only handwork in the hoophouse – no tractor or rototiller. The lack of rain indoors means there are very few weeds, and by not inverting the soil at all, we get few new weeds. We enjoy the peaceful, enclosed, intensively planted miniature garden, protected from winter weather and heated by the sun.

Persephone Days

If you want to harvest in the darkest days of winter, called Persephone Days by Eliot Coleman and the Solstice Slowdown by Alison and Paul Wiediger, you'll need to plan to have a good supply of mature crops to take you through. Be aware of the increase in days to maturity in winter. Here at latitude 38°N, daylength is less than 10 hours from Nov. 21 to Jan. 21; during those two months, there is little plant growth. For most of the winter, our plants are actively growing, not merely being stored for harvest, as happens farther north, so we can continue sowing new crops even in December.

Crop overview

Our winter crops are a mixture of direct sowings and transplants brought in from outside (in the fall), or grown inside and transplanted during the winter. We start in early September, sprouting some spinach seeds in a jar in the fridge, and sowing Bibb lettuces outdoors, to transplant later into the hoophouse. Meanwhile, we clear and compost one of the beds inside to sow the one-week-old sprouted spinach seed, some radishes, scallions, Bulls Blood beet greens and tatsoi. We make outdoor sowings on Sept. 15 and Sept. 24 for crops to later transplant into the hoophouse at 2-4 weeks old. The Sept. 15 sowing includes about 10 varieties of hardy leaf lettuce, Pak Choy, Chinese cabbage, Yukina Savoy, Tokyo Bekana, Maruba Santoh, and chard. The Sept. 24 sowing consists of Red and White Russian kales, another 10 varieties of lettuce, Senposai, more Yukina Savoy, Mizuna and arugula, and resows of anything from the previous week that didn't give a good stand of seedlings. We cover this outside bed with hoops and rowcover to keep bugs off, and water it frequently.

By the end of September we clear the summer crops from at least one more hoophouse bed, add compost and work it in. We transplant the Tokyo Bekana, and Maruba Santoh at just 2 weeks old, the Chinese cabbage, Pak Choy, Yukina Savoy at about 3 weeks old and the bibb lettuce at 4 weeks. (Plants grow so fast in September!)

By mid October we clear and prepare another bed and transplant 280 lettuce at 10" apart, and the chard. We also sow various tender turnips and radishes. We transplant the second lettuce (300 plants), the kale, Senposai, Mizuna and arugula and the second Yukina Savoy at about 4 weeks old, in the fourth week of October, after preparing the remaining beds.

We sow our first baby lettuce mix Oct. 24, along with our second spinach and chard. At the end of October, we sow some "filler lettuce" to transplant later to fill gaps. We try hard to keep all the space occupied, mostly using lettuce and spinach. On Nov. 10 we sow our second turnips, Mizuna and arugula, more filler lettuce and spinach, and also our first bulb onions for outdoor transplanting as early as possible in the new year – we aim for March 1.

From Nov. 10 on, we have a fully planted hoophouse, and as each crop harvest winds down, we replace that crop with another. Our winter and spring crops come to an end in March or early April, when we dig holes down the centers of the beds to plant the early summer crops of tomatoes, peppers, squash and cucumbers. Usually we harvest the winter crops from the center rows first, plant the new crop down the center, then harvest the outer rows bit by bit as the new crop needs the space or the light. This overlap allows the new crops to take over gradually.

Varieties that work for us

Asian Greens: We like the big heads of Prize Choy or Joi Choi Pak Choy, as well as Tokyo Bekana and Maruba Santoh, which have tender, bright yellow-green leaves. When small, the leaves are good in salad. When big,

The author's unheated hoophouse, above. Fava beans planted in November are ready for harvest the following May, below.

whole heads can be cut for quick-cooking greens. Tatsoi and its bigger cousin Yukina Savoy are good on flavor and strong dark color.

Chard: Bright Lights, Rainbow. Although green varieties grow faster in low light, we don't need a lot, and we enjoy the varied colors.

Chinese/Napa Cabbage: We like Blues, but the Michihili types are taller and narrower, and might make more sense in terms of space use.

Kale: the White and Red Russian kales do well in our hoophouse, but not outdoors, where they get weather-beaten. Inside, they do better than Vates, our outdoor kale.

Lettuce: Oscarde, Merlot, Tango, Red Tinged Winter, Lollo Divino, Galisse, Salad Bowl, Red Salad Bowl, Rouge d'Hiver, Winter Wonderland, Outredgeous, Hyper Red Rumpled Wave, Revolution, Green Forest Romaine. Looseleaf varieties, mostly. We look for ones that do well in winter, not just in early spring. Romaines and bibbs are good for harvest in December and late February onwards, but can get tipburn in midwinter.

Lettuce Mix: various mixtures, some homemade.

Mizuna: purple and green. Ruby Streaks looks very good. It adds texture to salad mixes.

Radishes: Cherry Belle are the "Industrial" radishes – all ready at once, quickly, looking perfect. We also like the multi-colored Easter Egg for variety and an extended picking season. It does well in the later plantings, staying mild in sometimes hot conditions in March, as does White Icicle, a long, crunchy, tender root.

Scallions: Evergreen Hardy White

Senposai: (the name translates as Thousand Wonder Green.) Fast-growing, tender, delicious. Grows big, and will provide bunches of leaves many times over.

Spinach: It's Tyee every time, for us: reliable, and tolerates varied conditions.

Turnips: Hakurei (short tops), Oasis, White Egg, Red Round.

Most of our seeds for greens come from Fedco, Johnny's or Southern Exposure.

Crops we tried, didn't like

Hot, bitter and sour greens don't go down well with our crowd. We have tried and rejected chicories, endives and pungent mustards. We're also less wowed by crops that are tiny or have to be harvested very promptly: Chinese broccoli (Hon Tsai Tai), mache (corn salad), claytonia. For us, it works better to use the space for more lettuce or spinach. In colder climates lettuce grows slowly, and smaller, hardier crops are more worthwhile.

We tried Nero di Toscano/Dinosaur kale, but it didn't do well for us. It harbored aphids in the curled leaves, and never developed the good flavor we hoped for. We have also tried various large Asian greens and hybrids that ultimately seemed too rambunctious for our hoophouse: Komatsuna, Mizspoona, Tyfon Holland greens.

Daily tasks

We reckon on about 2 hours work each day in winter, (only one hour in the summer). We aim to keep the temperature in the 65-80°F range, opening the big high windows, and some of the doors as needed.

If the sun is shining, we usually open the windows around 9 a.m. and close them around 2.30 p.m. to store some of the warmth. The number of daylight hours will depend on your latitude. On cloudy days, we use the thermometer to help us decide if we can open the windows. Even in cold weather, plants need fresh air! While photosynthesizing, plants need carbon dioxide, so just walk around in there, breathing, and you are doing "carbon dioxide enhancement"! When it's below freezing and cloudy, you'll want to keep the place closed up.

High-density cropping can really use up the carbon dioxide in a closed hoophouse very quickly. When this happens, photosynthesis crashes, and plant growth becomes limited. Good organic matter has high levels of soil organisms producing carbon dioxide, and dense plant canopies can trap this near soil level, where it is of great use.

Once the night temperatures are reliably above 45°F, we leave the windows open at night, to slow down the bolting tendencies of the cool weather overwintered annuals. In March and April we have to balance this need with that of the newly planted tomatoes, peppers, squash and cucumbers for warmth. If night temperatures will be above 55°F, we also leave the doors open.

Our main task each day is harvesting. In the winter of 2009-10, we had frozen soil or snow on the ground outside for a month (very unusual here). We were able to keep 100 people in fresh salad and cooking greens (with turnips and scallions for variety) for the whole month.

Aside from harvesting, jobs include planting new crops, clearing old ones, spreading compost, hoeing, hand-weeding, and supplying water as needed. We have drip irrigation, with 3 lines of drip tape in a bed. We hand water new sowings and transplants using a hose. In the middle of winter, not much water is needed, and we try to only water when a relatively mild night is forecast, with the idea of minimizing cell wall damage by not having turgid leaves in freezing conditions. For new hoophouse growers, it can come as a shock to find everything frozen in the morning, and equally shocking, but much more pleasant, to find everything thaws out later and recovers just fine.

Last winter's trial crops

Each year we try a few new ideas, or tweak something we tried the year before.

• Spring Cabbage Greens, sow outdoors Oct. 15, transplanted inside Nov. 2, harvested March 13-27. In the UK these are sown in early August, grown outdoors, and called spring greens. They are actually adolescent cabbages of particular varieties that will overwinter there. They are transplanted about 4-6" apart, harvested once by cutting alternate plants, then cleared as they start to head up. Some UK varieties include Offenham Flowers of Spring, Durham Early, Dorado, and Avon Crest. I've tried Early Jersey Wakefield, because they are the right pointed shape, but they don't work as well as the real thing. This crop provides a very dark green, great flavored unhearted cabbage. It took a bit of work to convince people to try them.

Celery grown outside for the summer, then transplanted big into the hoophouse Oct. 24, 12 plants, at 10-12". This keeps the plants alive, producing stems to add zest to winter stir-fries and salads (chopped). Harvested Dec. 26 - April 30.

Wild Garden Seeds' Pink Lettucy Mustard salad mix, sown Nov. 2. We're still experimenting to find the best sowing dates here.

Bulb onions sown Nov. 10, transplanted early March in a single row along the south edge of beds, for an early crop. We got good onions but they dwarfed the pepper plants behind them. Maybe planting them on the north side of the bed is better?

• Fava beans sown Nov. 15, harvested mid May. This worked well, we may do a bigger planting next winter. Legumes add diversity to the rotation.

• Snap Peas 'Sugar Ann' sown Feb.1, harvested April 21 - May 15. Very successful. We've now added this to our repertoire.

• Peashoots – sometimes a fallback crop (called a "catch crop" in the UK) if we have unexpected open space and leftover soaked seed from our spring outdoor planting, mid-March. Harvested April 10 - May 5.

• Brassica Salad Mix, sown Feb 6 and Feb. 12 for March and early April harvests. We mix our own from leftover random brassicas. For a single cut, almost all brassicas are suitable. Worthwhile if other crops fail, or outdoor conditions are dreadful and you need a quick crop to fill out what you have to sell.

Rotations

Rotation of crops in the hoophouse could be a big topic on its own, as the desire is to grow the same crops every year. Our approach is to vary crops, so that even if we end up with lettuce in the same bed two winters running, we have had two other crops there in between, such as squash and cowpeas. We do keep track and do our best to rotate, but the task is like making a patchwork quilt, where the scrap of Aunt Rosie's wedding dress just has to be included somewhere. Or Vegetable Sudoku, with only one possible solution.

Crop requirements

Because crops grow so fast in the hoophouse, the organic matter in the soil is consumed at a rapid rate. Each new crop requires a fertility boost. For fall planted crops, we spread 1" of compost across the beds and work it into the top of the soil with scuffle hoes or rakes. In general, for each crop, growers add compost at a rate of 15 tons/acre. We use about 4 wheelbarrows (not full) per 4' x 96' bed. A full wheelbarrow generally holds 6 cubic feet - about 40 gallons, or 170 liters.

For the early summer crops planted in March and April, we dig a hole for each plant and put a shovelful of compost in each hole. When we grow edamame or cowpeas in the middle of summer, we do not always add more compost at that point.

Transplanting

Because space is at a premium, it's good to measure plant spacings in some way, rather than drift to wider spacing and end up with fewer plants. We use wooden sticks. Rulers are cheap and convenient. Many growers turn on the irrigation, and use the wet spots to plant into. Hoophouse crops grow bigger than they would outdoors, so sometimes the spacing needs to be larger.

Cultivation

We use scuffle hoes, Cobrahead hoes, and the hand cultivators that are commonly available for backyard gardeners. We call them "Claws" – they have 3 tines and a short handle. Few weeds grow, but sometimes the soil texture looks in need of some help, because of the lack of rain.

Pests and diseases

Voles are our most frequent winter mammal visitor. We use mousetraps, the plastic Intruder type. Our worst winter insect pests are aphids and Vegetable Weevil Larvae. For the aphids, we bring in hibernating ladybugs (they wake up with warmer temperatures!) or spray with soap. For the Vegetable Weevil Larvae, which love turnips and Chinese cabbage, (but will move on to spinach), we use Spinosad. We also try to attract beneficial insects by planting flowers – you can put them in pots and move them around to where they are needed. We have used Alyssum and Echinacea and Buckwheat.

By far our worst disease problem is the fungus we call Solstice Slime. It's scientific name is Sclerotinia drop. It attacks lettuce when the days are short and temperatures chilly, causing the whole plant to collapse into a soggy grey pancake. So far, our approach has been to finish our shift by collecting up the slime and taking it outside, then washing our hands before returning. Prompt action does seem to limit the spread. The Wiedigers, who have several hoophouses, use solarization of one tunnel each summer to kill the spores.

Tipburn (brown leaf margins, including internal leaves) is caused by quick drying of the soil, when the weather makes a sudden switch to bright and sunny from overcast. Be ready to irrigate when the weather brightens.

Harvest

Starting in November, we harvest spinach, lettuce leaves, Mizuna, arugula, beet greens, Tatsoi, brassica mix for salad, as well as radishes and scallions. Lettuce is harvested by the leaf, leaving the center to keep growing. We switch to harvesting the heads in late January, when growth begins to pick up.

From December we also have baby lettuce mix, Tokyo Bekana, Maruba Santoh, chard, kale and turnips. Kale grows whenever it is above 40°F.

The bigger greens, including Senposai, Pak Choy, and Chinese cabbage, feed us from January till mid-March, if we plant enough. Yukina Savoy starts in the New Year.

Spinach is harvested by the leaf, using scissors, until the plants start to look a bit past their peak, then we "crew-cut" or buzz-cut them. Which brings me to another nice thing about hoophouse growing. For winter crops, (even kale), there is no need to remove finished plants to the compost pile, if they are not diseased. We simply leave them on the surface to dry up and disintegrate, improving the texture and nutrients in the surface layer of the soil.

Nitrate accumulation

In winter, when light levels are low, there is an increased likelihood of high levels of nitrates in leafy vegetables. This is generally thought to be a health hazard, (although recently this has been questioned) - nitrates can be converted in the body into nitrites, which reduce the blood's capacity to carry oxygen, and nitrites may be further converted into carcinogenic nitrosamines. To keep nitrate levels as low as possible:

- Grow varieties best suited for winter.
- Avoid fertilizing with blood meal or feather meal
- Water enough but not excessively.
- Provide fresh air so that carbon dioxide levels are high enough.
- Harvest after 4 or more hours of bright sunlight in winter.
- Avoid harvesting on very overcast days.
- Harvest crops a little under-mature, rather than over-mature.
- Use crops soon after harvest.
- Store harvested greens at temperatures close to freezing.
- Spinach contains about twice as much nitrate as lettuce, so mix your salads, don't just eat spinach.

Succession planting of winter hoophouse crops

Crop	Planting date	Harvest date	Notes
Chard	10/15 *	12/11-4/9	
	10/26	3/6-4/9	
Lettuce mix	10/24	12/11-2/21	
	2/1	3/20-4/20	3 cuts if lucky
Mizuna	10/24 *	11/1-1/25	
	11/10	1/27-3/6	
Onions, bulb	11/10		transplanted outdoors as early as possible in March
	11/22		
	12/6		
Radish	9/6	10/5 - 11/15	
	10/22	11/25 - 1/29	
	11/27	2/12 - 3/13	
	12/27	3/2 - 4/1	
	1/27	4/2 - 4/15	
Scallions	9/6	12/25 - 3/20	
	11/13	3/19 - 5/15	
Spinach	9/6	10/30 - 4/9	sown as sprouted seeds
	10/24	11/20 - 5/7	
	11/10		later plantings to fill gaps; harvested till 5/7
	12/27		
	1/17		
	1/24		to transplant outside later
Tatsoi	9/7	10/30 - 12/28	
	11/15	2/15 - 2/28	
Turnips	10/15	12/4 - 2/20	
	11/10	2/25 - 3/10	
	12/10	3/20	Only worthwhile if thinned promptly and eaten small. Greens are very sweet and beautiful in the hoophouse.
Yukina savoy	10/10 *	12/30 - 1/22	
	10/24	until 1/29	only one week extra

* First plantings that are transplanted rather than direct seeded.

Pam Dawling is the garden manager at Twin Oaks Community in central Virginia. Her book, Sustainable Market Farming: Intensive Vegetable Production on a Few Acres, is available at www.sustainablemarketfarming.com, or by mail order from Sustainable Market Farming, 138 Twin Oaks Road, Louisa, Virginia 23093. Enclose a check (payable to Twin Oaks) for $40.45 including shipping. Pam's blog is on her website and also on facebook.com/SustainableMarketFarming

Vegetables that can take the cold

By Pam Dawling

While some growers are eager to get their fields and gardens into winter cover crops, others see the last quarter of the year as an opportunity to sow and plant some cold-hardy vegetable crops. Some vegetables grow quickly and can be harvested before winter, others overwinter for early spring harvest, and some can even be harvested throughout the winter, depending on your climate. Because winter conditions can vary, success cannot be guaranteed. If a crop fails there is unlikely to be time to resow. I have plenty of respect for farmers who carefully guard their winter down-time in order to recover energy before next spring. And for those who plan to build barns or greenhouses, or fix equipment. If you don't have big plans and are not desperate for a rest, fall planting is worth considering among your options. As an extra plus, weeds grow slower in colder weather, planting dates may become more flexible, and you have fewer crops to take care of.

Winter Share CSAs will be much more in demand as the Locavore movement grows, as will year-round demand for local foods in stores and markets. After all, people eat year round, not just for the duration of a 26-week summer CSA. If you don't want to do the work yourself, perhaps you can rent out some land to a grower for the winter, or make some cooperative arrangement. There is a big potential for Winter Share vegetable CSAs to link up with Bread Shares, Herb Shares, and other value added products, strengthening the whole local economy. If you keep your customers through the winter, you'll have their loyalty next summer.

Before taking the plunge, know your climate, know your resources, know your market, know your crops, and when you don't know, experiment on a small scale.

Winter-hardy vegetables. Clockwise from left: Collards 'Morris Heading'; Kohlrabi; Senposai (a hybrid of cabbage and Japanese mustard spinach); Kale 'Vates'. Photos by Kathryn Simmons.

Your climate

Useful data include first frost date, daylight hours (related to latitude north or south of the equator), average annual minimum temperature (USDA Cold Hardiness Zone), and wind chill effects. Temperature is a big factor in the rate of crop growth, but it is not the only one. Excess soil moisture, wind strength, rodent and deer pressure also need consideration. Three hours of direct sun per day is needed — in winter the sun arcs lower in the sky than it does in the summer, possibly not making it above the horizon for as long as you think, if you live among hills. On the plus side, deciduous trees will be leafless, so you may get more sunlight than in summer!

Your resources

If you have a hoophouse, you'll be able to produce an abundance of fresh crops. If you don't have a hoophouse, fall is a good time to put one up. You can sow a nursery bed of transplants outdoors under row cover to give yourself a head start, then transplant the seedlings into your hoophouse once you get the plastic on. (You'll also get the chance to experience an interesting "season reversal" in weeds if you cover your hoophouse in late fall - first the cool weather weeds germinate, then the warm weather weeds return as the space warms up — a harbinger of what you can hope for in vegetables, too!) Whenever it is sunny, even if it's very cold outside, your winter hoophouse crops will be actively growing in the extra warmth.

Cold frames are useful. Heavy duty row cover with hoops and something to hold down the edges is also a valuable tool. Each layer of thick row cover adds 6F degrees of frost protection (but reduces light transmission to 75%). Standard weight row cover provides 4F protection. Depending on your climate, some crops may not need row covers, or may only need covering for the coldest nights. Kale, for instance, may only need to be covered for a short time during the winter to keep it alive for spring, when it can become very productive again.

Raised beds can help keep plants warmer and drier. Seaweed foliar sprays can increase the cold hardiness of your plants. Straw, tree leaves, pine straw or hay mulch around (not over) the plants can also help keep roots warmer.

What about your human resources? If your crew has slimmed down to yourself only, the workload will need to reflect this.

Your crops

Winter vegetable crops fall into three categories: those that grow fast and mature before the coldest weather, those that can be harvested all winter, and those that sit in the ground and get off to an early start in the spring, allowing an earlier harvest. Look for especially cold-tolerant varieties or gene lines. Brett Grohsgal and Christine Bergmark of Even' Star Farm and Frank Morton of Wild Garden Seeds are working to produce extra-hardy varieties. Consider investing in cold-tolerant seed stock and then saving your own seed to select for cold tolerance as it is needed at your farm. Sometimes heirloom strains are more cold tolerant than modern hybrids – winter hardiness has fallen out of favor in the era of highway-travelling veggies.

Fast fall crops

Look for cool-weather crops that mature in 60 days or less. Mostly these are greens and fast-growing root vegetables. Possibilities include salad greens (lettuce, endives, chicories), radishes (both the very fast small ones and the larger winter ones), and arugula. Also many kinds of oriental mustard greens: Chinese Napa cabbage, Komatsuna, Maruba Santoh, Mizuna, Pak Choy, Senposai, tatsoi, Tokyo Bekana, and Yukina Savoy. Spinach, kale, beets, chard, collards, kohlrabi, turnips, and small fast cabbage (Farao or Early Jersey Wakefield) all fall in this category.

If you use row cover to get through your first frost of the season, there is often a spell of warmer weather before more frosts (here at least), effectively pushing back your first frost date. This may allow a late succession sowing of warm weather crops such as bush beans, squash, or even cucumbers.

It is possible to reduce the Fall Factor if you are providing protection. Pre-sprouting seeds will reduce the number of days from seed to transplant.

Here's an example: Early White Vienna Kohlrabi takes 58 days from sowing to harvest (line 1). You can direct sow, so line 2 = 0. You can harvest it all at once and store it in your cooler, so line 3 is 1 day. Assuming you don't want to use row cover for this, line 4 = 14. Line 5 = 14 also. That all adds up to 87 days. Kohlrabi is hardy to maybe 15°F (-9.4°C). When is the temperature likely to drop to 15°F (-9.4°C)? Not before the beginning of November here, so counting back 31 days in October, plus 30 in September, plus 31 in August – that's 92 days already, more than enough. We could sow kohlrabi in early August and get a crop at the end of October.

While doing these calculations, consider sowing a winter cover crop such as winter rye between the crop rows a month before your harvest date. By then the cash crop will be big enough to withstand competition from the cover crop. This way, growing a late cash crop won't cause you to be too late to sow your winter cover crop.

Last possible sowing date for outdoors

For crops to be harvested before killing conditions arrive, the formula to find the last workable sowing date outside is:

Number of days from outdoor planting to harvest

+ Number of days from seed to transplant if growing your own starts

+ Average number of days of harvest period

+ 14 days "Fall Factor" to allow for the slowing rate of growth as the weather cools

+ 14 days "Safety Margin" from the date of killing temperatures for that crop

= Days to count back from the date of killing temperatures for that crop.

Winter harvest crops

For crops to harvest all winter long, look for those that will survive your lowest temperatures, taking any row covers or other crop protection into account. Consider arugula, some cabbage varieties, carrots in some areas, chard, collards, garlic scallions, horseradish, Jerusalem artichokes, kale, leeks, some onion scallions, parsley, parsnips, salsify, spinach, and very hardy salad crops (corn salad being the hardiest). Fall is too late to plant the slower-growing winter hardy crops like parsnips or leeks.

In the middle of winter, when there is less than 10 hours of daylight, little plant growth happens. At our farm, these dates are November 20-January 20. Because the really cold weather arrives later than the shortest day, our period of little growth is more like mid-December to the end of January. If you want to harvest during that period, sow extra so you can harvest just part of the patch each time. Spinach is the hardiest green we grow over winter (we like Tyee). To grow big enough to harvest in the winter, we sow this spinach at the very beginning of September. We use row covers on hoops to protect the leaves from the worst of the weather, and spinach will make new growth whenever the temperature rises above 37°F. We grow 7 beds, and use a marker flag to indicate which bed to pick next. If we arrive to harvest and find the spinach in the flagged bed isn't big enough, we wait.

If you get heavy snow-falls, a cold frame will be easier to deal with than row covers.

Overwintering crops

If you can keep certain crops alive through the winter, they will start to grow again with the least hint of spring weather and be harvestable earlier than spring plantings. These include spinach, kale, collards, some onion scallions, garlic scallions, chives, chard and arugula.

Spinach, lettuce, fennel and cilantro seem to have the best cold tolerance when the plants go into winter half-grown. We usually have one or two beds of spinach which we sow in late September, overwinter as adolescents and harvest in the spring. These plants bolt later than the ones we harvest leaves from all winter, and earlier than spring sown beds, so we get continuity of supply. Notice that fall sowing dates are quite exacting: Sept. 5 is the latest we can sow spinach for harvesting October-early April, and Sept. 20-30 sowings will not get big enough to harvest until late February.

Some cabbage (Deadon, Brunswick and January King) can survive in zone 7 for January harvest, but don't qualify as a fall-planted vegetable, as they are slow-growing and need to be started in summer.

With alliums such as bulb onions, multiplier onions and garlic, the harvest dates are regulated by day length, so the harvest will not be earlier but the bulbs will be bigger if you can over-winter them.

Winter hardiness

Record how well your crops do in the colder season. My friend and neighboring grower Ken Bezilla has suggested the morbidly named Death Bed idea: set aside a small bed and plant a few of each plant in it to audition for winter hardiness. Note when the various plants die of cold, to fine tune your planting for next year (and send me an email!) It's worth noting that in a hoophouse plants seem able to tolerate lower temperatures than those listed here: they have the pleasant daytime conditions in which to recover. All greens would do a lot better with some row cover to protect them against cold drying winds.

Here are some starting numbers, although your own experience with your soils, microclimates and rain levels may lead you to use different temperatures:

35°F: Basil.

32°F: Bush beans, cauliflower curds, corn, cowpeas, cucumbers, eggplant, limas, melons, okra, some Pak Choy, peanuts, peppers, potato vines, squash vines, sweet potato vines, tomatoes.

25°F: Broccoli heads, chervil, Chinese Napa cabbage (Blues), dill, endive, annual fennel, large leaves of lettuce (protected hearts and small plants will survive even colder temperatures), Maruba Santoh, mizuna, onion scallions, most Pak Choy, radicchio, Tokyo Bekana.

22°F: Arugula, tatsoi. (both may survive colder than this.)

20°F: Some beets, cabbage heads (the insides may still be good), celeriac, celtuce (stem lettuce), perhaps fennel, some mustards/oriental greens (Tendergreen, Tyfon Holland greens), flat leaf parsley, radishes, turnips (if they have leaves to protect them).

15°F: Some beets (Albina Verduna, Lutz Winterkeeper), beet leaves, broccoli leaves, young cabbage, celery (Ventura) with row cover, cilantro, endive, fava beans (Aquadulce Claudia), Russian kales, kohlrabi, Komatsuna, some lettuce, especially small and medium-sized plants (Marvel of 4 Seasons, Rouge d'Hiver, Winter Density), curly leaf parsley, oriental winter radish with leaves for protection (including daikon), large leaves of broad leaf sorrel, turnip leaves, winter cress.

12°F: Some cabbage (January King, Savoy types), carrots (Danvers, Oxheart), most collards, some fava beans (not the best-flavored ones), garlic tops if fairly large, most fall or summer varieties of leeks (Lincoln, King Richard), large tops of potato onions, Senposai.

10°F: Chard (Green chard is hardier than multi-colored types), mature cabbage, some collards (Morris Heading), some endive (Perfect, President), young stalks of Bronze fennel, perhaps Komatsuna, some leeks (American Flag), Oriental winter radish, (including daikon), rutabagas, (if they still have leaves to protect them), large leaves of spinach.

5°F: Garlic tops if still small, some leeks (Bulgarian Giant, Laura, Tadorna), some bulb onions, potato onions and other multiplier onions, smaller leaves of savoyed spinach (more hardy than flat leafed varieties) and broad leaf sorrel.

0°F: Chives, some collards (Blue Max, Winner), corn salad, garlic, horseradish, Jerusalem artichokes, Vates kale (although some leaves may be too damaged to use), a few leeks (Alaska, Durabel); some onion scallions (Evergreen Winter Hardy White, White Lisbon), parsnips, salad burnet, salsify, some spinach (Bloomsdale Savoy, Olympia, Tyee).

Keeping crops alive

If you want to overwinter crops, start by believing it is possible, and look for the hardier cultivars available. If you can add some wind protection, do so, especially for large leafy plants. At our zone 7 farm, we overwinter Vates kale without row covers, but we've killed off Winterbor and the Russian kales that way, while we were learning. Many greens and roots can survive some freezing, so it is worth experimenting to find how late you can keep crops outdoors. Don't harvest frozen crops – wait till they thaw.

Hoops can be used to keep row covers from sticking to frozen leaves and reduce abrasion. In winter we use double hoops – the outer hoops trap the rowcover, so it doesn't blow away. Ventilate covered crops in mild weather, so they don't lose their cold tolerance. The microclimate under hooped row covers is very pleasant in chilly, windy weather.

Brett Grohsgal of Even' Star Organic Farm in Maryland has done extensive research and plant breeding to create extra hardy greens. Brett and his family grow large fields of greens and root crops in the open, completely unprotected. They have started to sell seed of some cultivars for growers who want to do their own seed growing and develop locally suited lines. Their varieties will work for areas which get at least 3 hours direct sunlight each day, and have minimum winter temperatures of 0°F. Some of their lines are available through Fedco Seeds and Southern Exposure Seed Exchange.

Other considerations

The economics of season extension can be considered by using partial budgeting. Calculate the additional costs of using a particular kind of season extension for your chosen crop or group of crops, and the likely added income. Calculate the total change in projected income and decide if it is worthwhile.

Success with season extension is a matter of finding the balance point at which the time, money and energy you put in are still definitely worthwhile. The farther we go from a plant's natural season, the more costly the efforts to keep it alive and productive. I often imagine the pineapple growers in Victorian England – low-paid workers crafting thick straw ropes to wrap around tender plants for insulation, and stoking coal stoves day and night to grow a pineapple or two for the rich people in the Big House. Not a model to copy!

Money is not the only factor. A longer harvest season helps you retain and satisfy customers. Just how important this is will depend on your market: if students are your main customers, you may as well take a winter break (and a long summer break) too. But if you cater to snowbirds, then winter crops will be important. Season extension can help provide year round employment for your crew, which helps you retain skilled workers. The flip side is that you get less free time which might cut too far into your need for R&R. The ecological impact of fossil fuels to manufacture agricultural fabrics, and plastics disposal are worth thinking about. However, people are going to eat, and if it isn't you providing local sustainably-grown produce, then someone else is trucking it in, and the ecological costs of the shipping are usually higher than careful use of plastics.

Scheduling your crops

Earlier I gave a formula for calculating sowing dates of crops to harvest before winter. Another approach is this one from the Heirloom Gardener by Jeffrey Goss:

10 weeks before the average first frost date (Aug. 1 for us): beets, cabbage (Brunswick and January King for January harvest), daikon, leeks, winter lettuce (Marvel of 4 Seasons, Rouge d'Hiver, Winter Density), turnips, rutabagas, watercress

8 weeks before frost (Aug. 15): winter radish, fall spinach

7 weeks before frost (Aug. 22): kale

5 weeks before frost (Sept. 7): spring spinach

3 weeks before frost (Sept. 21): fall globe radishes

Eliot Coleman's *Four Season Harvest* (available from www.growingformarket.com/store) has a table of fall planting dates for various first frost dates. Use the columns with a first frost date 20-30 days later than yours for cold frame sowings, and the columns for 40 or more days later for hoophouse sowings. Using this, in September here with a first frost date of October 20, in USDA zone 7 (the temperature gets down to 5°F at some point), we can sow corn salad, claytonia, lettuce, endive, escarole, mizuna, onion scallions, radicchio, radish and spinach outdoors. In cold frames we can sow those crops into late October. In September we can still sow beets, early broccoli, Chinese cabbage, carrots, chard, chicory, kale, kohlrabi, leeks, mizuna, bulb onions, parsley, peas, sorrel and turnips in the cold frame. In the hoophouse, we could also sow cabbage, celery, celeriac, rutabagas, tatsoi in September.

Winter harvest techniques

With fall sown crops, often the aim is to keep the same plants alive all through the winter, as November, December and January are not good times to sow replacements. With leafy vegetables, the highest productivity is achieved with Cut and Come Again crops – the tops of the plants above the growing point are cut with scissors or shears every 10-35 days. Leaf by Leaf is the method we use for kale, collards, chard, and spinach (as well as outdoor lettuce under row cover). Never remove more than 40% of the total leaf surface area of the plant – less than half of the leaves, with a safety margin. Whole Plant Harvesting works well for small plants like tatsoi and corn salad. A direct seeded row can be thinned over time by harvesting out the biggest plants on each visit.

Pam Dawling is the garden manager at Twin Oaks Community in central Virginia. Her book, Sustainable Market Farming: Intensive Vegetable Production on a Few Acres, is available at www.sustainablemarketfarming.com, or by mail order from Sustainable Market Farming, 138 Twin Oaks Road, Louisa, Virginia 23093. Enclose a check (payable to Twin Oaks) for $40.45 including shipping. Pam's blog is on her website and also on facebook.com/SustainableMarketFarming

Crops to direct seed in fall

By Adam Montri

By mid-August, only seven of 20 beds in our 34 X 96 foot high tunnel have summer crops left in them. We'll leave the sweet potatoes for another month before we dig and cure them for winter farmers markets. The two beds of colored peppers will come out in the first or second week of October to make way for the final planting of baby salad. The last bed of eggplant probably should have been pulled already but we left them to ripen the last few fruits on each plant.

Planting the fall hoophouse is always a juggling act. You don't want to take out productive summer hoophouse crops any sooner than necessary. But you do need to plant fall crops in time to have them growing well before the days get too short.

At our farm, we tend to direct seed cool-season crops more than most of the farmers we know. There are a number of reasons for this, but the main one is that we both have multiple off-farm jobs so if we can direct seed something and water it once a day while it's germinating (or miss a day if we are out of town for work) we usually go for it. It also saves us time and money that we would spend buying or making media, filling flats, watering twice a day, etc. That shouldn't be seen as a knock on transplanting fall crops in tunnels or growing transplants on the farm. Many of our friends do that and we may be there someday, but, for now, this is the decision we have made for our farm based on our off-farm responsibilities.

Direct seeding, though, means we have to start fall planting earlier. For other farmers, transplanting fall crops allows the summer crops to stay in the tunnels longer — which means higher yield of the tomatoes, peppers, melons, and other summer crops that customers love.

If you haven't yet torn out your summer hoophouse crops, you still may have time to direct seed fall crops. In this article, I'll describe our direct seeding schedule, spacing, and seed costs for fall hoophouse crops.

Scheduling

The 13 beds that don't have summer crops are either already planted with fall crops, or they are being watered to flush the weeds before we plant. Beets seeded on August 13 are off and running after emerging in just three

Fall/Winter Hoophouse Crops — Spacing, Seed Use, and Seed Costs

Crop	Cultivar	Spacing (inches)	Seeds/ 28' Row	# of Rows	Avg Seeds/lb	Seed Packet Size and Cost	Seed Cost/ Bed
Asian Greens	Various	8 X 8	126 (3 seeds per space)	4	200,000	1/4 lb = $14	$0.14
Baby Salad		2.25 X 1	336	18			
	Garrison				384,000	1 oz = $19.50	$4.91
	Sulu				384,000	1 oz = $19.50	$4.91
	Mizuna				200,000	1/4 lb = $11	$1.33
	Red Russian Kale				112,000	1/4 lb = $15	$3.23
	Arugula - Astro				236,800	1/4 lb = $9	$0.91
	Red Giant Mustard				249,600	1/4 lb = $16	$1.55
Beets	Red Ace	7 X 2	168	5	35,000	5,000 = $11.55	$1.94
Carrots	Napoli	4.5 X 2	168	9	288,000	10,000 = $27.00 (pellete	$4.08
Collard Greens	Champion	12 X 12	84	3	112,000	1/4 lb = $7	$0.06
Head Lettuce	Various	8 X 8	126	4			
	Adriana				384,000	1/4 oz = $20.00	$1.68
	Winter Density				384,000	1 oz = $10.50	$0.22
Kale	Toscano	12 X 12	84	3	112,000	1 oz = $10.00	$0.36
Radish	Various	4.5 X 2	168	8			
	Crunchy Royale				40,000	5,000 = $6.50	$1.75
	D'Avignon				40,000	5,000 = $8.50	$2.28
Scallion	Evergreen Hardy White	2.25 X 1	336	18	200,000	1/4 lb = $10.50	$1.27
Spinach	Various	8 X 8	168	3			
	Space				35,000	5,000 = $5.50	$0.55
	Regiment				35,000	5,000 = $4.50	$0.45
	Covair				35,000	5,000 = $4.50	$0.45
Swiss Chard	Bright Lights	12 X 12	84	3	30,000	1,000 = $4.30	$1.08
Turnips	Various	4.5 X 4	84	8			
	Hakurei				185,000	1 oz = $19.00	$1.10
	Scarlett Queen				124,000	1 oz = $18.00	$1.56

days; carrots seeded the same day came up in four days in nice straight rows and are pushing their first true leaves; Swiss chard and Tuscano kale are coming up in small clumps of multiple plants soon to be thinned. Weeds are popping up in the open beds, too, leading to visions of stirrup hoes slicing them off like a guillotine, followed by weed-free beds for our three salad mixes. We usually don't have empty beds like we do this year, but we added a 26 X 48 foot movable tunnel in April. This house has our tomatoes in it and will have them again next year in a different location. Because of this we are finding ourselves with more open space in the big tunnel. If we were to use transplants, we could do a quick crop of radishes, turnips, or other fast crop but we are actually enjoying the chance to flush the weeds.

Our fall hoophouse plantings are a little different than our spring ones in that we tend to do more hand seeding in fall. This isn't because we want to torture ourselves by holding seeds in our hands while we scoot along close to the ground, trying not to tip over and drop all those seeds in the aisle or the bed next to us. It more has to do with getting the right spacing as we transition out of outdoor production. In the spring, we seed kale, chard, head lettuce, and other leafy or heading crops in the hoophouse using the Earthway seeder, which tends to put the seeds every 2 inches or so, depending on the plate. In the spring we use that as a way to produce our own field transplants directly in the hoophouse beds. Crops that are seeded in February and early March are ready to be transplanted out in late March and April. At that point, we transplant out most of the crops from the beds and leave them at either 8 X 8 inch spacing (head lettuce) or 12 X 12 inch spacing (kale, chard, other leafy greens). It's an easy and efficient way to grow our own transplants without a lot of extra work or cost. But in the fall, we don't need transplants for the field, so we seed in the hoophouse at the desired final spacing.

While our outdoor plantings stay relatively the same in spacing and density, very few crops are planted at the same spacing or density from year to year in the tunnels. They aren't significantly different, but there are little tweaks. For direct seeding in the tunnel, we use a six-row precision seeder and an Earthway seeder. We have moved over to seeding some seeds by hand (as mentioned above), but those are only in the fall and only for a few crops. We still transplant warm-season crops and buy them from a friend who has started a business around supplying transplants for hoophouse farmers that want them earlier and later in the year than they are otherwise available. It's really important to say again that we understand that transplanting fall and early spring leafy crops in the tunnels is a better use of time in them but this is what has worked for our farm.

Heat-loving crops in Alaska

A hoophouse allows Allison Lentz of Homer, Alaska, to grow artichokes, sweet corn, tomatoes, dahlias, and other summer crops. With the hoophouse's protection in spring, many plants will quickly catch up in the extremely long days of summer at northern latitudes.

Bed preparation

When we are growing crops year-round we get three or four crops per year in each bed. If we choose a quick crop like radishes or hakurei turnips we usually get four. Otherwise it's three. Between crops we use a Troy-bilt Horse that we can get down the outside aisles and maneuver in the beds without tilling or stepping on the beds directly next to what we are prepping. We used a broadfork, three-tine claw, and rake in the first two years of the farm and really like what that does for the beds, but have moved to the tiller because of how quickly it can prep the beds. We still use the fork-claw-rake once a year to break up any hard-pan that could form using the tiller each time. This happens between the summer and fall/winter crops. Even though it would be easier and we would have more time available if we did it in the spring, we think it is best at this time because we have been in there harvesting more and walking on the edges of the beds. When we have had a tomato or cucumber crop centered in a bed that we have been harvesting multiple times a week for many weeks we can definitely tell that the edges of the beds get stomped down and the aisles get much wider. This is in contrast to when we have a crop like salad mix or spinach in a bed that we are harvesting from the aisles and that takes up the entire plantable area of a bed so that we are not ever walking on the edges. Each 3 X 28 foot bed then gets 8 cubic feet of compost that is lightly worked in using the three-tine claw. We usually smooth the bed out with a rake and then plant. If we are using the six-row pinpoint seeder, we like to roll the bed once with no seed in it to firm it up and make it more level. If we are using the Earthway or seeding by hand we don't roll the bed.

Seed spacing and density

This is something that changes a lot for us. We started out with extremely close, tight spacings and have been moving away from them for most crops, except salad mix (which we grow a lot of) and scallions (which we don't grow anymore because scallions don't sell in our markets). At first we thought it was all about how closely we could get crops but have been discovering over the past three years that we think it is more about a balance between spacing, crop size, and disease management.

Carrots are the best example of this. When we started, we used to plant carrots at 18 rows in a 36-inch bed. There are farms that do this and grow great carrots, but for us we spent a lot of time picking through small carrots to get ones that we would put in a nice bunch of 12 and charge $3 for. We sold small baby carrot bunches but it was still time consuming to bunch all of those small carrots. Last year we moved to 12 rows and, of course, ended up with larger carrots but still felt that there were too many small ones we had to pick through (we also had to eat a lot of stew in the winter to use those small ones up). This year we have 9 rows of carrots in a 36 inch bed and from the looks of the spacing right now we are going to be very happy with this. The other advantage of the wider spacing is that it is easier to weed. Outside, we plant four rows of carrots in the same size bed so we are still more than double our outside density. Wider spacings also help to manage disease. The winter tunnels usually run around 50-65°F on a somewhat sunny day. We vent to try and lower the humidity but it is still fairly warm and very humid most days. By decreasing our planting density, we are hoping to see less disease through increased airflow, especially in the short, cold, cloudy days of late December and most of January.

On the next page is a table of crops that we have grown in the tunnels in the winter. We don't necessarily grow all of them anymore. The spacings we feel are a good starting point to finding the balance between intensive production, crop size at harvest, and disease management. If you are transplanting these crops, the spacings for the heading and leafy crops are the same as if they were direct seeded, and they can be planted later in the year. Also included are the costs per seed or per planting. This is a piece of our ongoing attempt to put together some enterprise budgets that are crop specific for tunnel production. We are recording labor, labor by activity, and a few other categories to really look closely at the economics of our farm. As these get recorded and entered we plan to have more tables like this available and potentially a more complete enterprise budget for winter production.

Adam Montri owns Ten Hens Farm in Bath, MI, and works with farmers through his outreach position with the Michigan State University Student Organic Farm. For more information, visit www.hoophouse.msu.edu.

Scheduling crops and revenue

The chart on the next page shows a planting plan and income projection for Adam and Dru Montri's high tunnel near Lansing, Michigan. It corresponds with the hoophouse pictured here, in which beds are laid out perpendicular to the length of the tunnel. The tunnel is 34 x 96 feet, with a 2-foot path on each side. Beds are 3x30 feet.

"We run them that way mostly because of the number of crops and the ability to then keep a relatively straightforward rotation plan in our heads by treating each bed more or less as a 'field,'" Adam said. "If we were growing more volume of a lesser number of crops, it would work for us to run the beds lengthwise."

High Tunnel Income Projection (Spring) — Ten Hens Farm — Adam and Dru Montri

Bed #	Crop	Plant Date	Sqft (beds = 90)	Projected Yield	Projected Gross Income	Notes	Assumptions
1	Carrots	30-Jan	90	120 Bunches	$360	Napoli	Bunches of 12
2	Carrots	21-Feb	90	120 Bunches	$360	Napoli	
3	Carrots	7-Mar	90	120 Bunches	$360	Napoli	
4	Beets	30-Jan	90	75 Bunches	$225	1/2 Red, 1/2 Golden	Bunches of 6
5	Swiss Chard	30-Jan	67.5 (3/4 Bed)	200 Bunches (40 plants)	$600	Bright Lights (bunches/plant)	
	Parsley	30-Jan	22.5 (1/4 Bed)	100 Bunches (20 plants)	$200		
6	Waido	15-Mar	45	7.5 lbs/cutting*	$293	Baby Salad with Bed 12	3 cuttings, $13/lb
	Red Russian Kale	15-Mar	45	7.5 lbs/cutting*	$293	Baby Salad with Bed 12	3 cuttings, $13/lb
7	Arugula	15-Mar	45	7.5 lbs/cutting*	$293	Baby Salad with Bed 12	3 cuttings, $13/lb
	Red Mustard	15-Mar	45	7.5 lbs/cutting*	$293	Baby Salad with Bed 12	3 cuttings, $13/lb
8	Hakurei Turnips	30-Jan	45	30 Bunches	$60	6/Bunch	
	Hakurei Turnips	1-Mar	45	30 Bunches	$60	6/Bunch	
9	Beets	1-Mar	90	75 Bunches	$225	1/2 Red, 1/2 Golden	Bunches of 6
10	Bibb Lettuce	30-Jan	22.5 (1/4 Bed)	27 Heads	$54	Ermosa	
	Romaine	30-Jan	22.5 (1/4 Bed)	27 Heads	$54	Jericho	
	Bibb	1-Mar	22.5 (1/4 Bed)	27 Heads	$54	Ermosa	
	Romaine	1-Mar	22.5 (1/4 Bed)	27 Heads	$54	Jericho	
11	White Scallions	30-Jan	45	100 Bunches	$200	Evergreen Hardy White	Bunches of 7
	Purple Scallions	30-Jan	45	100 Bunches	$200	Deep Purple	
12	Red Lettuce	1-Mar	45	7.5 lbs/cutting*	$293	Baby Salad with Beds 6 & 7	3 cuttings, $13/lb
	Green Lettuce	1-Mar	45	7.5 lbs/cutting*	$293	Baby Salad with Beds 6 & 7	3 cuttings, $13/lb
13	Red Lettuce	30-Jan	45	7.5 lbs/cutting*	$293	Baby Salad with Beds 14 & 15	3 cuttings, $13/lb
	Green Lettuce	30-Jan	45	7.5 lbs/cutting*	$293	Baby Salad with Beds 14 & 15	3 cuttings, $13/lb
14	Waido	21-Feb	45	7.5 lbs/cutting*	$293	Baby Salad with Beds 13 & 15	3 cuttings, $13/lb
	Red Russian Kale	21-Feb	45	7.5 lbs/cutting*	$293	Baby Salad with Beds 13 & 15	3 cuttings, $13/lb
15	Arugula	21-Feb	45	7.5 lbs/cutting*	$293	Baby Salad with Beds 13 & 14	3 cuttings, $13/lb
	Red Mustard	21-Feb	45	7.5 lbs/cutting*	$293	Baby Salad with Beds 13 & 14	3 cuttings, $13/lb
16	Dino Kale (full leaf)	30-Jan	45	150 Bunches (30 plants)	$300		
	Red Russian Kale (full leaf)	30-Jan	45	150 Bunches (30 plants)	$300		
17	Spinach (baby)	Fall	90	15 lbs/cutting	$450	3 cuttings @ $10/lb	
18	Spinach (full leaf)	30-Jan	90	15 lbs/picking	$450	3 pickings @ 10/lb	
19	Shelling Peas	30-Jan	30	0	$0	Double Row Down Middle	Family Use
	Easter Egg Radish	15-Mar	15	18 bunches	$36	Outside 1ft of Bed	Bunches of 12
	Watermelon Radish	15-Mar	15	18 bunches	$36	Outside 1ft of Bed	Bunches of 12
20	Snow Peas	30-Jan	30	?	?	Double Row Down Middle	Yields to base on?
	Easter Egg Radish	30-Jan	15	18 bunches	$36	Outside 1ft of Bed	Bunches of 12
	Watermelon Radish	30-Jan	15	18 bunches	$36	Outside 1ft of Bed	Bunches of 12
					$8,220		

* =Baby Salad Avg Yield When Components Combined = 1 lb/6 sqft/cutting; Root Crops assume 80% of potential yield;

Growing year-round in Kentucky

By Alison and Paul Wiediger

When market gardeners in most of the U.S. think about production, they think spring-summer-fall. We have found that one of the most productive and enjoyable times to garden is in the winter, November through February. We produce leaf lettuces, spinach, mesclun mix, red Russian kale, mustards, mizuna, arugula, beet greens, and more, all winter in an unheated hoophouse. We've also found several advantages to using the hoophouses in spring, summer and fall. We would like to share the details of our experiences with other market growers who want to use unheated hoophouses in their production schedule.

The structure

Our farm is located in Zone 6b, about 100 miles south of Louisville. We sell produce grown without the use of chemical fertilizers or pesticides in Bowling Green, Kentucky, a city of about 55,000. We have five steel Quonset style tunnel structures with double layers of inflated poly. The first was picked up at auction when a large discount store built a "superstore" and auctioned all items at the old store. It is a 21x60-foot steel structure and we paid $750 for it. Serendipity! (It has since been expanded to 90 feet long). We have one more like that and the other three are what many greenhouse manufacturers used to sell as cold frames for wintering over perennials. We paid $740 for a 20x96-foot frame in 1998. Prices have changed significantly, and you still have to add site preparation, endwalls, toe and/or hip boards, door(s), irrigation, poly, an inflation fan and more. We spent just under $2,000 or about $1 per square foot. Prices have changed, but the value of a hoophouse for a market farm has not.

We think the double poly layer makes the system work for year-round growing. The inner layer is 6 mil IR (infrared). The outer layer is standard 6 mil, 4-year poly. It provides several degrees of protection for the plants inside. Even in cloudy weather, we often have a gain of 10 to 15 degrees. In sunny weather, the interior temperature will soar above 80 degrees, even when the temperature outside is below 20 degrees. The soil temperature rarely falls below 50 degrees and we have earthworm activity all winter. At night, the air temperature will hold as much as 10 degrees above the outside temperature at a height of 5 feet, and at plant level, temperatures are even warmer.

The Wiedigers' hoophouse in fall, with crops showing good growth before the dark days of winter set in.

Bed layout

The inside of the hoophouse is divided into five permanent beds with 18-inch pathways between them. The beds on the outside walls are 24 inches, which allows us to reach all the way across them. The other three beds are 42 inches across. We have chosen 18-inch aisles because a narrower aisle doesn't allow us to work and harvest without damaging plants. By keeping permanent beds and paths, we rarely need anything other than hand tools to get the beds prepared for the next crop unless there is heavy crop residue to turn under or weeds get away from us.

These beds, because they are used intensively, are amended twice a year – before the spring-summer season and before the fall-winter season. At that time, we add any compost we have, about 1 inch deep, and composted chicken manure (Replenish 3-4-3) as well as any crop residues. The amendments are spread on the top and worked in with a spading fork or wheel hoe with scuffle hoe attachments and in some instances, a rear tine tiller. Beds are smoothed with a rake and ready to plant.

Aside from fertility, moisture control seems to be the most important part of the equation. We need to keep the soil moist enough for steady plant growth and not so wet it encourages disease. Some years, we can have the soil at a good moisture level in late November and not need to water until February. However, most years, we need to water more to keep adequate moisture for good crop production.

To keep from chilling the soil and plants, when we need to water, we watch the weather forecast for a window of several sunny days and nights above freezing. We water on the first day of that window, so that by the time the freezing temperatures return, the plants have adjusted to the extra water and the soil has warmed. We use drip irrigation, but hand watering does work if you have the time or prefer it.

Seed or transplant?

We have found that we get the best continued production if we transplant plugs into the beds rather than trying to direct seed. The plug transplants are raised outside for the first fall planting, then in our heated transplant greenhouse for later plantings. We usually use 96-cell inserts so we have a robust transplant and plan on about five weeks from seeding to transplant. True plug trays, say 288-cell, also work very well if you have the time and discipline to keep them moist.

We space our lettuce transplants about 1 foot apart each way in the bed to allow the plants room for nice head development. At this spacing, the transplants make a canopy over the soil quickly, which slows weed growth. We transplant enough lettuce plants every three to four weeks during the season to supply our current market for a three to four week sales window. This requires seeding the flats every three to four weeks as well. Although we use our heated greenhouse for this task, we think it could also be done on our small scale with grow lights. A five-week transplant usually gives us a nice head of lettuce in four weeks from date of transplant except for the shortest days in December and January.

We have done some trialing of direct seeding and found that for mesclun mix, it can take as few as 21-25 days when we seeded in November but seeding in January can take twice as long. However, with regular cutting, the early sowings kept producing.

We have grown many varieties of lettuce over the years. We found that the Grand Rapids and oak leaf types were the best producers in the winter but we also got nice heads from green Bibbs and Romaines. The red lettuces don't keep their color well and don't grow as well in the low-light conditions of winter.

The varieties of lettuce that produce best for us are: Waldmann's (dark green), Vulcan (a superior red leaf), and Merlot or Galactic (a deep wine red but small). We also grow Pinot and Drunken Woman Frizzy Headed, which sells like crazy because of its name! For Oakleaf we plant Panisse (dense, compact plant with lime-green leaves), and Oscarde (dense, compact plant similar to Panisse but with red leaves). We also grow a number of Bibb or Buttercrunch lettuces, including Buttercrunch (dark green), or Nancy (medium green), Red Cross (red) or Pirat (red) and Deer Tongue (a small heirloom Butterhead, with a buttery yet crisp texture). Our favorite Romaines are Kalura, Winter Density (small), Rouge d'Hiver (red French heirloom) and Freckles (mottled colored).

We grow spinach from direct seeding. Although we have tried many varieties, we still use Space, a smooth leafed variety. It handles the temperatures really well, is easy to wash, and makes a large plant. We grow beet greens from direct seeding. They are a beautiful addition to salads, but get a little burned in the lowest temperatures. However, they will outgrow that damage when the temperatures moderate.

We direct seed Arugula, Red and White Russian Kale, Red Bor Kale, Endive, Green and Red Mustard, Mizuna and a leaf lettuce mix (reds and greens) for mesclun. Through early December, we can cut once every five to seven days, depending upon light levels - it takes a little longer when it is overcast.

Many other types of greens do well in the hoophouse. Some that we have grown are Chinese Cabbage, Kales, Collards, and Swiss Chard. Also, don't forget the sweetest carrots ever, salad turnips, radishes that last for weeks and don't get hot, and many other crops. Your imagination will inspire you and your market will lead you.

Early tomatoes

About the third week in February, the late fall transplanted lettuce starts to get a little bitter and the other greens start to bolt. At that time, we harvest the remaining spinach and freeze it for summer quiches, cut the rest of the greens and feed them to the cattle (they LOVE that late winter shot of vitamins!) and prepare the planting beds. About the 15th of March, we set tomato transplants into the beds. Yes, the hoophouse is still unheated and we are lower USDA zone 6 with an "official" last frost date of May 7th.

In 1998 we had a 10-degree freeze on the 10th of March and we set out tomatoes in the hoophouse on the 16th; in 1999 we had snow and ice on the 13th and set on the 16th. We usually have lengths of Remay cut to cover them, just in case, but we have had to use it only a handful of times over the years. These tomato transplants are grown in 36-cell, deep cell inserts and are about 8 weeks old when we set them.

The tomatoes are set 22 inches apart, down the center of the 42" bed. We also set lettuce transplants down each side of the bed in front of the tomatoes. The lettuce is harvested before the tomatoes need the space. We use 4 lines of drip tape for irrigation, one on each side of the tomato row, and one more on each side for the lettuce. Steel fence posts between every two plants and Florida Weave work well for supporting the tomato plants.

We use Bush Early Girl for taste, fruit size and determinate habit. We want an 8-10 oz. tomato for farmers market, and this variety gives us that, and some fruits much larger. After using indeterminate types for several years in the beginning, we choose determinates as the easiest to manage in this environment, and by the time they are finished, our field tomatoes are ripening. We also plant a few yellow and red cherries for market appeal – it really draws the customers in! And Bush Champion for a slightly later red slicer. We usually have ripe tomatoes to sell by the end of May - about four weeks earlier than field-grown tomatoes. That gives us a great price at the market, and a great market share.

Summer salad mix

I'd bet many of you are like us: by the time you have good ripe tomatoes and cool cucumbers for salads, the heat has "done the lettuce in." Our customers love our lettuce, and we have been able to offer it to them most of the year. But mid-June through mid-September, lettuce hasn't been "growable" here in Kentucky. We tried for years to successfully grow sweet, tasty mesclun during the hot months. We finally got it right. It takes a hoophouse to get that clean baby lettuce. You may or may not need shade. We have grown it both ways. But by paying attention to timing we can grow mesclun that will rival the quality of our fall/winter/spring offerings.

One change we make is to take the greens out of the mix. We use lettuces only. The greens get a little too much insect damage, mostly flea beetle, to make a nice mix. And, if you DO use shade, it may take some experimenting to get good reds. We found they "greened out" under shade cloth.

We seed the mesclun about 17 days before the anticipated harvest date for June through August plantings. Before and after that, it will need about 21 days. We only cut it once, and then turn under the residue. That helps cut down on insects and disease. It sounds like a lot of seeding, but we have done the math: it takes about 3 ounces to seed 300 row feet, and we can cut at least 10 pounds of mesclun (at $9 per pound) from that seeding. Seed cost: $10.

It will need a lot of water—plan on using the drip at least once a week, maybe twice if it's really hot. We sometimes have to use BT spray in late summer to control caterpillars, but the harvest is well worth the effort, even with yields that are lighter than in cool weather.

Summer into fall

By July, the tomatoes have seen their best and the field tomatoes are bearing, so the tomatoes come out and go to the compost heap. The beds are amended and reshaped. Depending on your market, this can be a great time to transplant melon plants. Melons do well here outdoors, but they do love heat, and in the summer, the hoophouse is HOT. Just keep them well watered and they will grow like crazy! This can give you melons for the late summer/fall market, after the melons planted outdoors are gone. Just be aware that coyotes and foxes will like them as much as you, and may come into your hoophouse after the fragrant ripe melons.

Once the melons are done, it's time to prepare the beds for the fall/winter season. Here, we can put the earliest transplants in about the first of September, although we are often late because of the pressure of harvest time in our outdoor growing. If we get them in late (some years it has been as late as November 1), they still grow and we still have our 'salads in the snow'.

In conclusion, we consider the hoophouse to be an essential part of our market garden. The structure can literally be used 12 months of the year in much of the temperate area of this country. The hardest part is getting some sort of a structure in place. It is so healthy, both physically and emotionally. There is nothing better than bundling up on a cold morning, then walking inside to 70+ degrees and picking a salad for your meal, or $300+ worth of produce for delivery. It picks up our spirits to the point we don't dread winter anymore. We can always walk to 'spring' and it doesn't cost a plane ticket! Try it! You'll never make a better investment.

Alison and Paul Wiediger own Au Naturel Farm in Smiths Grove Kentucky. They have written a booklet about their hoophouse production, Walking to Spring. *To order and for more information on their farm, visit their web site: http://www.aunaturelfarm.com*

Even California winters benefit from hoophouses

By Deborah Walton

In California, we don't have snow. (At least, not where people are farming – we keep it in the mountains where it belongs.) And we don't have very many days below freezing. But we still have weather in the winter – wet, cold rains that turn fields into bogs, splash dirt on every leaf in sight, and generally send small farmers scurrying into their warm, snug homes for two or three months.

At our farm, Canvas Ranch, we grow veggies year-round. We are 50 miles north of San Francisco, right on the border between Marin and Sonoma counties. We are six miles from the coast in a valley that gets a lot of fog in the summer, and about 25 inches of rain a year. (Unless there's a drought, as there always seems to be.) In winter, we can grow broccoli, cauliflower, kale, collards, chard, and Brussels sprouts in the field because they like a touch of frost. (They didn't like the heavy frost we had in December, however.) For some of the crops, we use floating row covers to add a few more degrees of warmth as well.

We also wanted to grow carrots, lettuce, sorrel, bok choy, mustard greens, spinach, and a few other yummy things that were a bit too delicate for winter field production. So several years ago we decided to erect a hoophouse.

We tried a simple, homemade PVC structure the first year. We spent most of the day constructing it . . . and the next morning it was over in the neighbor's yard. The "neighbor" in our case is the U.S. Coast Guard, a branch of Homeland Security, which didn't really appreciate the big plastic thing flying over their fence!

Husband Tim said, "Enough of that," so we ordered a 30 x 50 foot kit hoophouse from an agricultural supply company, which cost about $3,000. It is made of steel pipes instead of PVC, and is tall enough to drive our tractor inside to till the soil. It took about a week to erect and put the plastic on. We used 6 mil greenhouse film rated to last for four years.

It can be 40-50°F outside during the day, and 70-80°F inside the hoophouse if the sun is out. The solar power generated in there is amazing. It also makes farming in the winter so much more pleasant, the plants happy and healthy, and CSA customers (and their farmer) smile.

We grow a lot of different crops all winter long, usually started in the greenhouse and put into the hoophouse as transplants to give them a jump start. In the spring, we'll transition to spring and summer crops, adding sugar snap peas, beets, radishes, and cherry tomatoes, along with more of the leafy greens. This way, if we can't get into the fields because of late rains, we still have crops producing.

I've had a few failures in the hoophouse. I tried twice to grow sweet potatoes, but got a nasty spider mite infestation one year, and dry, crinkly, brown plants the second. Not sure what the problem was with either of those, but I've given up trying to push the climate that much.

Our other big problem is gophers, who love the warmth that the hoophouse provides. Cinch traps and a few hoophouse "spa nights" for our cats help a bit, but they are still a major nuisance.

Following Eliot Coleman's model, the hoophouse is laid out with seven 3x50-foot beds with 18" paths between them. A wider center path of 2 feet allows the farmers to easily get to the far end to open the window for ventilation. A light layer of wood chips helps with the mud. Below: Shelves for hardening off flats of transplants line one wall of the unheated hoophouse. The shelves are attached with hinges so they can be folded up when not needed. Lettuces are grown underneath, so they also protect the tender greens from intense sunlight. Photos by Deborah Walton.

But all that pales when you are harvesting sweet greens, tender scallions, and beautiful beets all year long. We have had a waiting list for our winter CSA every year, attracting people who in the summer head to the farmers market or grow their own backyard vegetables. It has been so successful that we're putting up another hoophouse next year.

Deborah Walton owns Canvas Ranch, www.canvasranch.com, in Petaluma, California.

In warm climates, hoophouses open new markets, pose special challenges

Northern growers led the way in the use of hoophouses for season extension, but the Noble Foundation in Ardmore, Oklahoma, has not been far behind. For the past decade, Steve Upson has trialed a wide variety of crops at the Foundation's farm in southern Oklahoma to see how hoophouses can best be used in the South.

Season extension is the chief benefit of hoophouse production in his climate, just as it is in the North, Steve says.

"Four weeks earliness in spring tomato production offers the same competitive advantage to

Oklahoma growers as four weeks earliness does for Kansas growers," he said. "I emphasize greater profit potential as a result of higher prices associated with non-traditional market windows."

The hoophouse's ability to protect crops from storms is another important benefit Steve has identified. "A crop is only worth as much as the protection provided by a 6-mil sheet of plastic," he says. "During the fall of 1996, our hoophouse tomato crop escaped the ravages of a hailstorm accompanied by 70 mph wind. Seeing is believing."

Besides protecting crops from storms, the hoophouses have provided other benefits:

•Foliar disease is reduced because crops are sheltered from rainfall.

•Some degree of freeze protection is provided, even without heat. Tomatoes in the hoophouse survived a night of 27°F outside temperatures.

•Crow and mockingbird damage, which occurs every year on field tomatoes, doesn't happen in the hoophouse even when the sidewalls are rolled up. Crows don't venture inside. Still, it's a good idea to put shade cloth on the sidewalls to act as a windbreak as well as keep out vertebrates.

Heat buildup

In the warm South, managing the temperature in the hoophouse can be a challenge, Steve says. Many factors affect heat gain, and here are some suggestions he has for dealing with them:

•The bigger the house, the greater the heat gain. Larger houses will require ventilating earlier in the day than smaller houses. Houses should be no more than 20 feet wide to allow for sufficient crossventilation.

•Clear greenhouse film coverings heat up fast. Steve prefers to use "Kool-Lite Plus" brand, which is treated to block more infrared rays than conventional films. Although it costs about 75% more, Steve says the Kool-Lite Plus film can lower hoophouse temperatures up to 12 degrees F during peak afternoon hours. "In my opinion, the additional cost can be justified considering the costs associated with the use of shade fabric, not to mention the difference in crop quality," Steve says.

•Shade fabric has mixed results. It can lower temperature a few degrees, but it also reduces the amount of light available for photosynthesis. Figuring out the amount of shade and timing of installing it is also a problem. Steve uses 55% shade on any house with regular clear poly.

•The quonset style hoophouse traps heat above the roll-up sides, which is good in spring and fall, but bad once the weather gets hot. Recently, the Noble Foundation built two new structures that Steve calls triple side vent hoophouses. This design has straight walls, one four feet higher than the other. The low side has one roll-up vent, and the high side has two, one at ground level and another at the top of the sidewall. The cost was about twice that of a conventional hoophouse, and Steve will be evaluating the new houses to determine if their cost can be justified.

Because temperature management is so difficult in summer, Steve says, they generally avoid using the houses those months. "We do, however, take advantage of the heat. Our philosophy is if you can't beat the heat use it to your advantage. Thus we cover the beds in our houses with clear plastic mulch and close them up. Closing the houses makes soil solarization more efficient. It is possible to generate temperatures as high as 130° F. at a oneinch depth in the hoophouse beds. The level of soilborne disease control is exceptional."

Tomatoes

Steve has been pleased with tomato production in the hoophouse. One house, 20x68 feet, yielded 3,162 pounds, which if sold for $1.50 a pound would bring a net return of about $2,900.

That budget was based on the variety 'Celebrity', an All-American selection that is high yielding, and high quality, with exceptional flavor. It is not an early maturing variety, though. And hoophouse growers are looking for earliness more than anything else with tomatoes, so they can get the higher prices the market offers before local field tomatoes come in. Earliness is also important in a hot climate because summer temperatures in the hoophouse may get too hot for tomato flowering, putting an early end to the crop.

Steve is on a quest to find the perfect hoophouse tomato for his climate. Recently, he trialed six early varieties, with Celebrity as the control, to find a good-tasting, high yielding, big early tomato.

The variety that yielded the heaviest early was 'Bush Early Girl', but it was small. 'Sunstart' and 'Sunshine' were slightly less prolific, but they were both big and early. 'Merced' had half-pound fruits, earlier than Celebrity but not as early as Sunstart and Sunshine. Steve plans to continue to trial new varieties as they are released to find the perfect hoophouse tomato.

The budget for the hoophouse tomato project is reprinted on the next page. You'll notice that the prices used in the budget analysis are very low, ranging from 50 cents to $1.50 per pound. Many direct market farmers get prices ranging from $2 to $4 a pound retail.

Tomato/Spinach

In 2007 through 2010, The Noble Foundation evaluated a year-round cropping system consisting of cool-season spinach and warm-season tomato production.

Growing began in August with soil solarization, followed by planting 'Melody' spinach in September. Spinach was harvested November through February. In March, 'Mountain Spring' tomatoes were planted and harvested in May, June, and July.

The researchers applied a traditional crop enterprise budget to the yield and expense data they collected over three growing seasons. Their goal was to determine the price growers would need to charge for spinach and tomatoes to be profitable. They concluded that the cropping system was viable. "Results suggest that growers can expect minimal production risk from year to year due to relatively consistent crop yields," they wrote. "Annul profitability for the spinach/tomato cropping system depends on factors such as the inclusion of the value of the owner/operator labor in total cost of production, the quantity of each crop not marketed (either through household consumption or post-harvest waste), and the consumers' willingness to pay..." Given the average marketable yield, and assuming the farmer includes his or her wages in the calculations, the break-even price for spinach was $3.04 a pound and for tomatoes was 86 cents per pound.

The researchers noted that both crops were labor-intensive, spinach being more so than tomatoes. Labor for spinach was 80 hours per 1,400-square-foot hoophouse and labor for tomatoes was 72 hours per hoophouse. Harvest accounted for the greatest percentage of labor (44%) for spinach, and crop care accounted for the greatest percentage of labor (38%) for tomatoes. Planting was the least labor-intensive chore, accounting for only 3% of total labor for each crop.

Peppers

Another crop well-suited to the hoophouse, because it likes heat, is the bell pepper. But can colored peppers be grown in the hoophouse? It takes 45-60 days from the time a pepper flower is pollinated until the fruit reaches its maximum green size. It takes an additional three weeks for the color to develop.

Pepper plants were transplanted in the hoophouse on March 24. Harvest began on June 23 and concluded July 6, by which time high temperature had reduced quality. None of the varieties developed a uniform mature color, but people at the Noble Foundation considered the multicolored looks appealing. Of nine varieties trialed, 'Mandarin' had the highest marketable yield and highest fruit weight.

Cucumbers

The cucumber variety 'Dasher-II' was planted in the hoophouse on March 24. On May 5, the first crop was harvested. Final harvest was June 9. A total of 3,727 marketable fruit was harvested from the 256 plants in the 1,360 square foot house.

"In our study, 90 man-hours of labor was required to grow the crop," Steve reported. "If the owners or manager perform tasks, net income will be significant. If they take a hands-off approach and hire everything done, profits will be reduced to a few hundred dollars."

Cut flowers

Annual flowers have been one of the most profitable crops at the Noble Foundation farm. Transplants of 12 varieties in 4-inch pots were planted through black plastic mulch in the hoophouse on March 24. Harvest began on April 8 and continued through June 26. Prices for the cut flowers were quoted by a local florist and ranged from 5 cents a stem for cosmos 'Bright Lights' to 75 cents a stem for sunflowers. Steve calculated the value of each variety as though each 20x68 house had been full of it. Gross income ranged from $2,352 for the cosmos to $18,720 for ageratum 'Red Top' (which was almost twice as productive as ageratum 'Blue Horizon').

Other noteworthy varieties in the trial: Calendula 'Prince'; Celosia 'Century'; Marigold 'Gold Coin'; Salvia 'Victoria'; Snapdragon 'Rocket'; Sunflower 'Valentine'; and Zinnia 'Ruffles'.

Details of these and other hoophouse research projects at the Noble Foundation can be found on its web site at http://www.noble.org

Hoophouse Tomato Budget

House size: 20 x 68 feet (1,360 square feet)

Crop frequency: One per year

Operating inputs				
Item	**Unit**	**Price**	**Quantity**	**Value**
Plants	4-inch transplant	$0.60	128	$76.80
Water	1,000 gallons	$1.66	14.4	$24.00
Pesticide				$10.00
Plastic mulch	linear foot	$0.04	250	$10.00
Fertilizer	pound	$0.50	20	$10.00
Compost	40# bag	$2.20	24	$52.80
Tiller rental	day	$45.00	1	$45.00
Labor	hour	$6.00	108	$648.00
Operating interest				$14.61

Total Operating Costs **$891.21**

Fixed Costs

Item	Amount	Depreciation (years)	Value
Building			
Structure	$3,027.00	10	$302.70
Windbreak curtains	298.00	10	29.80
Plastic film	404.00	3	134.60
Raised beds			
Lumber	251.00	10	25.10
Soil (16 yards)	246.00	10	24.60
Equipment			
Drip irrigation	483.50	10	48.35
Irrometer	50.00	10	5.00
Tomato cages	235.00	5	46.80
Cage support	88.00	10	8.80
Backpack sprayer	570.00	10	57.00
Interest Charge	682.75	10%	68.27
Land cost	.03 acres ($5,000/acre)		13.00

Total Fixed Costs **$764.02**

Total Costs (Operating and Fixed) **$1,655.23**

Production

Item	Unit	Price	Quantity	Value
Tomatoes	Pounds	$ 1.00	3162	$3,162.00
Total Costs (Operating and Fixed)				**$1,655.23**
Net Returns (Receipts - Costs)				**$1,506.77**

Annual Net Returns for Different Prices and Production of Hoophouse Tomatoes

Price received per pound

Pounds produced	$ 0.50	$ 0.75	$ 1.00	$ 1.25	$ 1.50
2000	– 655.23	– 155.23	344.77	844.77	1344.77
2250	– 530.23	32.27	32.27	1157.27	1719.77
2500	– 405.23	219.77	844.77	1469.77	2094.77
2750	– 280.23	407.27	1094.77	1782.27	2469.77
3000	– 155.23	594.77	1344.77	2094.77	2844.77
3250	– 30.23	782.27	1594.77	2407.27	3219.77
3500	94.77	969.77	1844.77	2719.77	3594.77

Source: Steve Upson, Reprinted with permission from The Samuel Roberts Noble Foundation, PO Box 2180, Ardmore, OK 73402. www.noble.org

The best cucumbers for the high tunnel

By Lynn Byczynski

Cucumbers can be a valuable crop when grown in a high tunnel. Like tomatoes, they are grown vertically on a trellis or string, so the yield per square foot can be impressive. The protected environment of the high tunnel allows for early and late production, and insect screening may thwart cucumber beetles in locations where the insects are a serious problem. Cucumbers can be grown in spring with tomatoes, or they can be the second crop after tomatoes, when tomato prices fall in mid summer.

There are many kinds and hundreds of varieties of cucumbers, though, and it can be daunting to figure out which ones merit the valuable space in the high tunnel. Lewis Jett, commercial vegetable crops specialist at West Virginia University, has been researching high tunnel cucumbers for many years and has both general and specific recommendations.

First, you need to know the terminology of cucumbers. Slicers produce fruits with seeds and a relatively thick skin, so they are good for both retail and wholesale. European, or Dutch cucumbers (also known as hothouse or greenhouse cucumbers) are parthenocarpic — that is, they don't require pollination to set fruit and are consequently seedless. They're also sometimes called burpless because they reportedly produce less gas in sensitive people than other types. European cucumbers also tend to have thin skin and they lose fresh weight rapidly after harvest, which is why they are often shrink-wrapped in the supermarket. Beit Alpha cucumbers are also seedless but have a waxy layer on the skin that prevents fresh weight loss.

In evaluating cucumber cultivars for high tunnel production, Jett has found that parthenocarpic cukes don't do as well in high tunnels as in greenhouses. In the closed environment of a greenhouse, they don't get pollinated so they set seedless fruits. But in a wide-open high tunnel in summer, bees and other pollinators are free to visit the cucumber flowers. When they do, some of the fruits develop seeds and a bottleneck shape.

Last summer, Jett tested 10 cucumber cultivars to find those that work best for high tunnel production in late summer. Seven were varieties currently available from Johnny's or Seedway and three were experimental. Seeds were planted in 50-cell trays in mid-August and 2-week-old transplants were planted in a high tunnel on Sept. 3. Plants were spaced 12 inches in the row, with 42 inches between rows. Each plant was pruned to one stem and trellised on a string trellis.

Harvest began on Oct. 10 and ran through Nov. 10, when a freeze killed the plants. Among the currently available varieties, the Beit Alpha 'Socrates' from Johnny's produced the largest number of marketable cucumbers per plant or per linear foot of row. (Other top performers were two of the experimental varieties.) In one month of harvesting, 'Socrates' (shown at left) produced 5.2 cukes per plant. 'Socrates' is noted for performing well in cool weather, so it's a good choice for fall crops.

Another high yielder was 'Rocky', a baby cucumber from Johnny's, producing 6.9 cukes per plant. 'Tasty Green' and 'Tyria' had more misshapen fruit as a percentage of total marketable yield. "Tyria" was seedless with excellent quality, but yield was not significantly high.

Jett noted that yields would have been much higher if the cucumbers had been started in July, possibly after an earlier crop of tomatoes. Cucumbers also can be grown in spring through summer in the high tunnel.

Previous research conducted by Jett in Missouri identified several other varieties as good choices for the hoophouse. The highest yielding varieties in a 2005 trial were 'Dasher II', a slicing type from Harris Seed; 'Indy', a slicer from Seedway; and 'Diva', a slicer from Johnny's.

Jett's 2011 research report is available at http://www.hort.purdue.edu/fruitveg/rep_pres/2011-12/toc11.shtml

His 2005 research report can be found at www.hightunnels.org/ForGrowers/

Galia melons recommended in Florida

A potentially profitable hoophouse crop is the delicious and expensive hybrid muskmelon Galia. University of Florida researchers found that Galia, which doesn't do well in Florida in the field, did much better in a hoophouse. Plants produced 5 melons each compared to 1 melon per plant in the field. Muskmelons must ripen on the vine to reach peak sugar content, and are therefore a great crop for direct marketing. Florida consumers pay $5-$6 for a Galia melon, the researchers said.

Annual strawberries a great crop

Strawberries can be grown as annuals in the hoophouse with high yields and great quality, research at the University of New Hampshire shows.

In seven years of trials, Dr. Bill Lord succeeded with the annual hill system of strawberry production, which is used in California and Florida. He found that the system, both in the hoophouse and in the field, produces bigger yields, higher quality fruit, and a longer harvest season than traditional matted-row production. Annual strawberries can be harvested a month earlier in the field and two months earlier in the hoophouse, when consumers are eager for fresh fruit and prices are higher. Berries can be grown on land that was in other crops the previous summer. And, best of all, annual hill strawberries require no weeding.

On the down side, production costs for the annual system are considerably higher. Lord suggests that annual strawberries are best for growers who want to get into the market earlier, or for those who have lost a matted-row planting to drought, weeds or disease the previous year but don't want to lose strawberry customers in spring.

The system

In California and Florida, strawberry plug plants are set into fumigated, clear-plastic-covered raised beds in late summer or early autumn. The plants grow through the warm winter and normally crop from midwinter to late spring. Elsewhere, strawberries are normally planted from bare-root plants in early spring. Flowers are pinched to prevent fruit formation the first year, and the beds must be weeded all year - a problem even for nonorganic growers, because of a lack of registered preemergent herbicides for strawberries. The first harvest comes the second year, in late spring. Here's how the annual system differs:

Preparing the soil. Raised beds are made with a bed shaper that also lays a line of drip tape and lays black plastic mulch. The beds are 6-7 inches tall and 24 inches wide. Plugs are set into the plastic, about a foot apart, in a staggered double row on either side of the drip tape.The black plastic is critical to the system for two reasons, Lord said: It controls weeds and keeps the soil warm well into fall to help the plants get established.

Planting. Strawberry plug plants are purchased from a supplier for fall planting. They're about twice the price of bare root plants, and need to be planted at twice the density of matted-row production, so plant costs are four times that of the traditional system. Several varieties have been trialed in New Hampshire and the best one, perhaps surprisingly, is the same one most used in California: 'Chandler.' A good second choice, Lord said, is 'Cavendish.' The main supplier Lord has used for plugs is Jersey Asparagus Farms (800-499-0013). The planting date varies by location. In Durham, New Hampshire, plants are set out on September 1. If the weather is unusually hot, planting will be delayed several days. Farther north, planting might be a week earlier. The key is to get the plants in the ground at least a month before the weather gets cold.

Mulch and row cover. Plants are foliar fed about the third week of September, or three weeks after planting. An organic mulch such as straw or hay is spread in the paths between beds to prevent soil from splashing up onto the plants and provide a clean environment for harvest in spring. In the field, in October, floating row cover - Typar 518 - is spread over the field.

"If there's a comment I hear most often from growers the first time," Lord said, "it's 'I can't believe those little plants are going to be able to produce any fruit in spring.'" But in the protected environment of black plastic mulch and row cover, the plants continue to grow as winter descends. Lord found that the row cover keeps the crowns of the plants five degrees warmer than a straw mulch. Plants that were 12°F under the row cover suffered no winter damage, but those under straw mulch at 7°F were not as productive. Row cover is removed when flowering begins in April, but kept on the side of the field in case frost threatens. "Growers do have to be vigilant about frost protection with this system," Lord said. Overhead irrigation may also be used to protect the crop from frost injury.

Harvest and yield. In Lord's trials, harvest began June 1, about two to three weeks ahead of traditional matted-row strawberries. "They tend to miss that hot blast of air that concentrates on the normal strawberry harvest," Lord said. "Often, we'll get to the first of July and have seven or eight days of temperatures in the 90s," which cuts short the strawberry season after two weeks. In comparison, the early strawberries fruited for four weeks in more pleasant weather.

Yields were much higher for the annual strawberries. The average yield from matted row plantings in the area is 8,000 pounds per acre; the annual strawberries produced 10,000 to 11,000 pounds per acre. Growers wondered whether those productive crowns could be left in the ground, so trial plantings were renovated by mowing and removing runner plants. Plants responded with excellent regrowth and were covered with Typar 518 on Oct. 1. Harvest the second year was a week or more later than from first year plants, probably because the bigger plants shaded the black plastic and reduced the soil warming effect. But yields were even higher than first year, about 12,000 pounds per acre. By then, the crowns had gotten too big, and the black plastic mulch was deteriorating, so the plantings were destroyed.

Because of the plastic mulch on the beds and straw mulch between the beds, the strawberries were clean and of excellent quality. The cull rate was just 15%.

Hoophouse production

The New Hampshire researchers took the annual system inside a high tunnel - and got all the benefits magnified. The tunnels they used were 14-foot-wide, 8-foot tall unheated plastic houses with roll-up sides. They planted the plugs a month later than in the field, on October 1, on the same size beds covered with black plastic mulch. The plants were not covered with row cover in the tunnel, and suffered no winter damage.

Harvest began on May 1, a month earlier than the annual field plants, and two months ahead of traditional outside matted row plants. Fruit size was excellent, averaging 20 grams per berry - equal to 21 or 22 berries per pound. Because foliage and flowers were never wet, no fruit rot damage occurred. And perhaps because the berries were so early, there was no damage from tarnished plant bug or clipper beetle. Yield was 30,000 pounds per acre, or about 400 pounds per 100-foot bed, planted in the staggered double row described above.

Lord continues to tweak the annual system in search of greater profits. For example, he used shade cloth on the tunnel from mid-March through April to extend the short-day environment strawberries need for flower initiation. More flowers means more fruits, and the plants in the shaded tunnel did in fact out-yield the others. Yield was 1.7 pounds of marketable fruit per plant in the shaded tunnel versus 1.4 pounds in the other. He also tried to trigger a second fruiting season by use of shade cloth. He was successful in getting the plants to fruit in July, but the plants suffered from heat, insect problems and a boron deficiency, and yield was poor.

This photo was taken in December at the author's farm. It shows a bed of strawberries, planted as plugs into black plastic in September. Whereas most of the other plants in the hoophouse have succumbed to frost, the strawberries are still green. The plants will not be covered with row cover, which may cause them to flower prematurely. In harsh winters, the leaves may die back but the crowns will be fine. The photo on page 5 shows the same bed of strawberries in April.

Autumn strawberries are possible

USDA researcher Fumiomi Takeda has demonstrated that strawberries can be coaxed into fall production in an unheated hoophouse with no loss of yield the following spring. But he is not aware of any commercial growers who have adopted his system to date. This seems like something that hoophouse growers might find worthwhile, at least on an experimental basis. If you are curious, here are the details you need to know.

First, you will have to produce your own transplants, which means you need to have access to appropriate varieties of mother plants, either from your own strawberry planting or another grower's field. Takeda had the greatest success with short-day varieties, especially Chandler and Carmine. The plants should be sending out runners that are not in contact with soil; for example, in a suspended growing system or pots in a greenhouse, or on plastic mulch in the field.

Takeda found that it's critical to take runner tips on July 1 for transplant production. He tried rooting the tips of Chandler strawberry plants on August 1 and found that none of the plants flowered in fall. When he plugged the tips on July 1, all of them flowered and fruited. Each plant produced about 3/4-lb. of berries from late October to the end of December. What's more, the transplants started July 1 also produced more fruit the following spring than those started later.

Strawberry transplants are easy to produce in a greenhouse, hoophouse, or shade house. Runner tips should be cut off when little white or brown pegs (called root initials) are visible and no bigger than 1/2 inch. Each tip needs to have at least two leaves between 2-1/2 and 4 inches in length. Any larger or smaller, and the plant may not root. About 1/2 inch of the runner should be removed along with the leaves, to anchor the plant. Tips should be planted immediately into 48- or 50-cell trays, with the root initials and anchor just below soil surface.

In mid December

'Carmine' strawberry, November 1, 2006

The flats of transplants should be placed under a misting system so the leaves can be misted every 5 minutes at first, gradually decreasing to about every 12 minutes as the plants start to root. A runner tip will produce a well-rooted plug in about four weeks; it should be left to grow in the plug tray for another month.

Takeda, whose research was conducted in the mid-Atlantic region, transplanted the strawberry plugs on September 1 into a plasticulture system in an unheated hoophouse. Raised beds were covered with black plastic mulch and water and fertilizers were provided by a drip irrigation system. When temperatures threatened to drop below 34°F, the plants were covered with 1.5 oz. row cover.

A publication about growing strawberry plugs from runner tips:
http://extension.usu.edu/files/publications/publication/Horticulture_Fruit_2010-01pr.pdf

At left, Dr. Takeda and Michael Newell, Horticultural Crops Manager at Wye Research and Education Center, University of Maryland, Queenstown, MD, show their strawberry production system. That year, the research was conducted in grow bags of perlite and fertigated 2-3 times a day. Since then, plants have been grown in raised beds with similar results.

Earliness, but no profit, from matted-row system

Doug Coldwell, a Massachusetts grower, grew three varieties of strawberries in the matted row system in high tunnels with funding from the Sustainable Agriculture Research and Education (SARE) program. Here is the summary the SARE staff wrote:

The three varieties were Annapolis, Kent, and Northeaster. He planted June 1 in raised beds and staggered double rows, and followed organic practices throughout. In August he sowed oats along the walkways for weed control, and to supply some of the mulch that would be required for winter cover. He applied mulch in November, removed the mulch the following March, and covered the greenhouses (which until this point had remained uncovered) with a double layer of poly film. He harvested strawberries from May 18 to June 15. His findings are as follows:

•One foot spacing within rows, and between the two rows of the double row, was a good planting density. He reports that the beds filled in nicely.

•The late planting date facilitated weed control.

•Sowing oats late in the establishment year worked well.

•The raised beds made cultivation with a rototiller more difficult, and improved drainage excessively on his light-textured soil. He recommends raised beds only for heavy soils.

•He recommends trickle irrigation.

•The high tunnels protected the berries from too much wetness, and saved the bulk of his crop. Field growers suffered severe losses to the excessive rains that June.

•Use of high tunnels enabled him to bring his berries to market 3.5 weeks before field-grown berries were ready in his area. He believes that covering the high tunnels earlier than March of the second year would have allowed him to bring his fruit to market even earlier. If this is to be done, however, the structure must be built to bear a load of snow.

•Regulating temperature inside the tunnels can be a problem.

•Doug used a whisk broom for pollination. He reports that this was too rough on the blossoms, and suggests a leafblower or bees, instead.

•The Annapolis variety yielded about half again as much as either Kent or Northeaster.

He received a wholesale price of $1.75 per pint. Although he considered it an excellent price, it was not enough to cover expenses. Yields would have to be higher, he says, to make this system of growing strawberries worthwhile.

Hoophouse raspberries for late-season profits

In their first year, berries produced all the way till Thanksgiving

By George DeVault

We first heard about planting raspberries in high tunnels at a New Jersey vegetable growers association conference in Atlantic City a few years ago.

The very idea of holding a farming conference in a gambling casino seemed absolutely ridiculous. While the setting was perhaps the perfect paradox, there were no oxymorons among the many seasoned speakers and the practical information they presented.

Several speakers talked about their experiences with growing raspberries in high tunnels. Tunnel-grown berries had a shelf life of up to 11 days, they said.

That sounded good, almost too good to be true, especially when such incredible news was delivered not far from the 4,100 slot machines and 200 gaming tables that fill the nearly 4-acre casino at the Borgata Hotel Casino & Spa.

Like most growers, we were used to growing berries outside, and then losing a lot of the crop to gray mold practically overnight. So, the chances of raspberries really lasting that long seemed just about as likely as us paying off our mortgage in one evening at the blackjack or roulette tables.

Still, the idea of everlasting raspberries lingered. Wouldn't they be nice? So last spring, we finally gave in to temptation and ordered 100 raspberry plants for our "annual experiment."

While waiting for the plants to arrive, I Googled "raspberries in high tunnels." In just 0.07 seconds, I had 455 authoritative references from all over the country and the world on my computer screen, just a mouse click away!

Researchers at Penn State waltzed us right back into the casino: "Raspberries should be considered a 'high stakes' crop.' Raspberries have a short shelf life and a short marketing season, but demand for raspberries is usually excellent and high prices can be obtained."

From Downeast, the headline on a 2005 Maine Organic Farmers and Gardeners newsletter also spoke seductively of risks -- and riches: "Raspberries: Challenging but Potentially Profitable ... Raspberries are one of the more challenging crops I deal with," says Maine's vegetable and small fruit specialist David Handley. "I have more people get started in and get out of raspberries than any other crop I deal with." People often see raspberries priced around $5 per half pint and plant the fruit. "But when the reality sets in as to how much labor is involved in getting the crop off the plant and to market in good shape, you see it's worth every penny of that $5, and the growers are earning their money. Lack of labor tends to be the downfall ... The market is certainly there."

Another Penn State bulletin (*Raspberries in High Tunnels: Opportunities and Challenges*) played it close to the vest: "When profits for high tunnel crops are compared, the big winner appears to be tomatoes -- raspberries fare well in the comparison, but still are in second place at best. Most growers report pay-off of the tunnel in either the second or third year with raspberry production. The big factors for long-term profits are yield, the price received, and the cost of labor for harvest. Good yields (10,000 lb/a equivalent or higher), a decent price received ($2.50 per half-pint or higher), and minimizing harvest costs ($0.50/half-pint or less) result in good potential for very significant profits."

Probably the two most complete and helpful references that came up were from Cornell: "High Tunnel Raspberries and Blackberries," a 32-page report revised in 2009, and the 38-page "Greenhouse Raspberry Production Guide" from 2003. I downloaded both for free.

Short high tunnels -- those that are only 7.5 feet tall -- are not recommended. "These tunnels are too low for raspberries as the house will get too hot. Installing post extensions of 4 to 5 ft. will greatly improve air circulation and reduce heating. Tunnels wider than 14 ft. are also more efficient for managing raspberries. Tunnels should be no wider than 30 ft. for good cross ventilation and reduction of snow accumulation on the roof in the winter. Tunnels longer than 96 ft. may pose a psychological barrier to pickers."

Armed with that sage advice and more, we prepared planting beds inside two of our tunnels and waited for our 100 plants to arrive. On April 20, 2010, we divided the plants evenly between the "Big Mama" and the "Brother," as we affectionately call the houses. Big Mama is our largest tunnel, a 30- by 96-foot structure. It was so named by our children because my wife, Melanie, stands a little under five foot two and is known as the "Little Mama." Brother is a 21- by 96-foot tunnel named in honor of our son, Don, who built it all — his way — all by himself, but that's another story. We bought the gothic arch frames from Ledgewood Farms in New Hampshire (www.ledgewood-farm.com). Both tunnels have roll-up sides and eight-foot square doors at each end for ventilation and easy access.

The variety we planted is Caroline, a patented ever-bearing raspberry that is said to have "fantastic ... rich, intense raspberry flavor." It is widely adapted (Zones 4-8) and matures five to 10 days earlier than Heritage. Fruit is described as "huge, very sweet, firm." Caroline is said to be more resistant to root rot than Heritage, with vigorous growth habits, disease resistance and exceptional fruit quality.

Our tunnels are oriented east-west. In each tunnel, we planted on the far northern side so that, when mature, the four- to five-foot tall raspberries would not shade lower-growing crops. (We goofed in Brother, and trellised heirloom tomatoes in the bed next to the raspberries. Shading from the tomatoes reduced raspberry yields a bit. In the Big Mama, the dahlias trellised beside raspberries didn't seem to bother the berries.) Using stakes and a string, we planted a single row in each tunnel, three feet in from the outside wall. Plants were spaced about two feet apart in the row. We squeezed some in closer together to create walkways at each end of the row.

For trellising, we simply tied three to five tiers of tomato twine to metal T-posts. The posts were sunk about one foot into the ground by hand using an inexpensive post driver. It didn't take long to figure out that five-foot posts are the safest bet under the sloping frame next to the outside wall. On the side toward the middle of the tunnel, there is no height limit and six-foot posts work just fine. The trellis is nothing fancy, expensive or even semi-permanent, but it does the job quite well.

With the berries under cover, irrigation was essential. I put two drip tapes with four-inch emitter spacing down in the Big Mama, but only one in Brother. On average, the Big Mama produced twice as many berries as the Brother. The problem wasn't just irrigation, however. For starters, soil in the Big Mama was better. I also tilled in some hog manure there, reclaiming the space taken up by an emergency hog pen our son had cobbled together to corral three hogs that kept getting through his electric fence. The Big Mama is covered by two layers of 10-year-old plastic that cuts a lot of light and almost acts as shade cloth. A single layer of much newer plastic covers the Brother, which is much hotter and brighter. We lost a lot of berries to solar damage (groups of individual druplets turning white and hard) in the Brother.

Harvest in both tunnels started in mid- to late August. It was a little hard to pinpoint an exact date, because most of the earliest berries were eaten right off of plant, wound up on our breakfast cereal or were served as a special snack for our then 16-month-old grandson, Forest DeLong.

Red raspberries, you see, are Melanie's absolute favorite fruit. And Mel felt like she had hit the jackpot at raspberry roulette. Besides eating raspberries right off of the cane, out of the box I was picking into or the walk-in cooler, every time she carried flowers into or out of the cooler, she:

* Put up 12 pints of raspberry jam.

* Froze 12 pints of raspberry sauce.

* Filled 15 freezer bags with from one to three cups of raspberries each.

Baked a few fresh raspberry pies, each of which requires 3 cups of raspberries

"You sold all the rest!" Mel said defensively when I asked her exactly how many berries we had put away for the winter.

Despite her hoarding, we still had plenty of raspberries to take to our Sunday farmers' market three miles away in Emmaus. Yields steadily increased through September. Picking took an hour or two, sometimes three, every day or two. And the berries stored extremely well in the cooler. We could start picking Sunday evening for market the next Sunday, and the berries would be just fine in both taste and appearance. All season, we lost only one berry — one individual berry, not a whole box — to gray mold.

Caroline certainly lived up to its reputation. The berries were large, firm and sweet, unlike the pretty but tasteless ones often found in supermarkets.

On Oct. 3, we took 30 half-pint boxes to market. Some weeks, we had about 45 boxes. But, again, exact numbers are a little hard to come by, since the contents of my berry flats kept changing by the hour due to gifts, grandmotherly treats and simple snacking.

I do know for sure that we took at least 30 boxes of raspberries to market on each of the five Sundays last October. Demand was so great that we put a limit of two boxes per customer on the berries. At $4 a box, we still sold out in the first 30 to 45 minutes.

Then November arrived, and the berries still kept coming. Customers were astounded, and even more eager to buy our berries. "How do you still have raspberries?" they would ask. "My berries at home quit weeks ago with the first hard frost!" Our high tunnel berries lasted all the way up to our last regular market on the Sunday before Thanksgiving.

While we didn't get rich on the high tunnel berries, they did add significantly to our weekly bottom line. More importantly, having raspberries at market when no one else did quickly conditioned customers to line up at our stand -- first. That greatly boosted our sales of cut flowers, vegetables, eggs and herbs.

This year promises to be even better, since our high tunnel berries are already established. To help them along, we'll prune and fertilizer the berries in both tunnels this spring. Then we'll add shade cloth and another drip line in the Brother. The two-box limit will be replaced by a 50 cent-a-box price increase.

At $4.50 a half-pint, high tunnel raspberries still look like a safe bet.

When he's not picking berries, George DeVault raises vegetables and pastured pigs in southeastern Pennsylvania.

Manage the hoophouse for more tomatoes, cukes

By Andrew Mefferd

As the snow falls here in Maine and I look back through my records for last year, I realize that somehow or another, I ended up having a good hoophouse year. In fact, I doubled my tomato yield from the previous year in the same size house. In 2011, I grew just over 2,000 pounds of tomatoes in my 30 x 48 foot hoophouse. In 2012, I grew just over 4,000 pounds. I'd like to share some of my hoophouse successes and failures from 2012 in the hope that I can pass on something useful for future crops. This is the first in a series of articles about hoophouse improvements, and will cover planting density and trellising methods.

Planting density

One of the most important tools that you can use to improve your hoophouse production is a pencil and paper. When confronted with a new idea of how to do things, it's important to "pencil it out," look at it on paper and see if it makes as much sense there as it does in your head. Such a change is what I give most of the credit for doubling my tomato yield. I have two 30 x 48 foot unheated hoophouses. At 1,440 square feet each, that's not much protected cropping space, and very valuable real estate to me. I grow tomatoes in one and cucumbers in the other, and switch back and forth every year. It's not the greatest rotation, but I figure it's better than tomatoes every year.

In 2011, I used the kind of spacing that I see a lot of people with hoophouses using — a single row of tomatoes with a head (the growing tip of the plant) every foot or so. I think of row spacing in the number of heads per foot rather than plants, because this expresses the correct spacing regardless of how many heads are on the plants. For example, if you wanted to have one head per foot over 50 feet, that would be 50 single leader plants, or 25 double leader plants, both producing the same head spacing. So for the purposes of this article, I'm going to talk in terms of number of heads to figure out plant spacing.

Previously in my 30 x 48 hoophouses, I was planting five rows of tomatoes, one head every foot, with rows 40 feet long. I am planting indeterminate, grafted, double leader tomato plants, so in my case those 40 heads were from 20 plants with two heads each. So at five rows with 40 heads each, that gave me 200 heads in the house.

In my job doing tomato variety research and development for Johnny's Selected Seeds, I get to visit a lot of greenhouses. I noticed that many of the larger greenhouses I was visiting use a double row of tomatoes, rather than a single, with the two rows 2 feet apart, and a 3-4 foot walkway between double rows.

I began thinking about the implications this spacing change could have on my own little hoophouse, so I got out my pencil, and after a little scribbling realized that I could accommodate four double rows of tomatoes two feet apart, with four feet between double rows in my 30 x 48 house. This raised the number of tomatoes I could squeeze in from 200 heads (1.25 plants per square meter) to 320 heads (2 plants per square meter), a 60 percent increase.

This increase in number of plants also results in a decrease in the amount of airflow. Although I did see a little bit more botrytis, in particular in the middle of the double rows, it wasn't that much worse than my old, lower density spacing. I just kept an eye out for disease outbreaks and removed infected leaves when I saw them. Though disease was a little higher I don't think it ever reached an economically damaging threshold and it never killed any plants.

I am pretty scrupulous about disease prevention, as far as scouting, removing old leaves below the developing fruit cluster, removing diseased foliage, sanitizing pruning and harvest tools regularly, and using rollup sides, peak vents, and HAF fans to improve ventilation. It's also worth noting that my hoophouses are relatively new and haven't had decades of use to build up a thriving population of diseases. Plus, our long, cold winters up here in Maine help the disease equation by killing off some pathogens that might overwinter in sunnier climes. So if there is a downfall of the high-density planting option, it's the potential for higher disease.

Planting day in May 2012 at One Drop Farm in Cornville, Maine. The farm's apprentices, Ben Hummel and Lindsay Glick, set out big, grafted transplants in a double-row pattern. Photo by Ann Mefferd.

The author in his hoophouse in 2011 with old spacing of five single rows. A lot of sunlight hitting the floor could be used to grow tomatoes. Photo by Ann Mefferd.

Two heads per square meter is still a low density by commercial greenhouse standards. But since I am growing in a hoophouse, not a greenhouse where I could control humidity, I am worried about foliar disease outbreaks if I get the spacing too dense.

I still think there is some under-utilized space in the house that I plan on taking advantage of next year. There is almost 5 feet from the outer crop rows to the edge of the hoophouse. The very edge of the hoophouse does not have the full height of the body of the house, because the slant of the roof is very close to meeting the sidewalls by that point. But, there is still a good six feet or more of height there. So I plan on running a purlin down each side of the house, a foot and a half from the sidewall, and trellising a single row of cherry tomatoes, or a short heirloom like Cosmonaut Volkov for example, down each edge of the hoophouse.

If I'm really feeling rowdy, I might plant a row of tomatoes along the south wall of the house, perpendicular to the other rows in the house. I have always been searching for a way to make better use of the space we use for walking at the ends of the house. Now the north end is no good, because it is shady from all the plants south of it. But the south wall is pure sun. I might have to set up a trellis structure inside the house, more like a field setup such as T-posts with a wire along the top, since there is nothing to trellis to a foot and a half from the south end wall. But I think that setup would officially maximize the holding capacity of my hoophouses.

But who knows? Maybe you can think of some way to sneak a few more plants in here or there? If you do, let me know. People have long used other creative strategies to get the most out of their houses. Hanging baskets of flowers or even cucumbers over a low-growing crop like seedlings or bedding plants is another strategy to use more vertical growing space. It's just not an option for me with 9-foot tall tomatoes or cukes in my houses.

Lowering and leaning

One reason beside plant density that double rows are standard in the greenhouse industry is because they use the two rows for lowering and leaning the plants as the vines grow. Lowering and leaning is a technique that enables growing a plant that is longer than the height of the structure that it is being grown in. The plant is grown up a piece of twine, with extra twine spooled and clipped securely to an overhead cable. There are several different devices available to attach the spool of twine to the overhead cable until it needs to be played out. The most common of these is called the Tomahook but there are others that work well, too. I use this system in my house because otherwise the plants would grow up to the top cable and run out of room sometime in late July or August, with some of the best growing yet to come.

When the plant has climbed the length of the twine, reaches the overhead cable and runs out of room to grow up any more, some of the twine is spooled out and the plant is lowered, giving it room to continue to grow up. The plant also needs to be leaned after lowering, or the stem would build up and possibly kink and break after repeated lowerings. So lowering is always accompanied by leaning, or pushing the plant down the wire to keep the stem from piling up at the base. The effect by the end of the season is that the head of the plant is many feet away from the roots, with the stem trailing along the ground where the plant has been leaned.

Since there is nowhere for the plants on the end to slide, they have to be looped around to the other row in the double-row pair. In this manner, the tomato plants are unlimited in the height they can achieve. Of course they are limited by the season and how far the roots can pump water to the head. But the vines can get very, very long using this method in a heated, long-season greenhouse.

The top wire in my hoophouse is 9 feet above the ground. We paid a little more for the extra high sides on our hoophouse, which is not that much more expensive and I think well worth the investment for anyone wanting to grow tall, trellised vining crops. I have never measured my plants at the end of the season, but I would guess they are approaching 20 feet long even in our short season and without heat. For the record, we plant our hoophouse tomatoes the first or second week of May and they usually get killed by frost in the first week of October or so.

I'd like to pause here and put in a plug for grafting. My plants are not much less vigorous the day they die than the day they get transplanted. The only way I know of to achieve this level of vigor is through grafting. I have seen hoophouses even in our short season area that lose vigor and stop pushing fruit before they reach the top wire and before the end of the season. If you have problems with low vigor in your tomatoes, it's well worth it to try some grafted plants.

I realize that in some parts of the country, the best time of year for hoophouse tomatoes may not be the summer, as it is here in Maine. Even with climate change here in the far northeast, our field tomato season is so short that sometimes I wonder how people even grow tomatoes outdoors here. By the time they are ripening, it is cooling off again. Which is why I love my hoophouse. It gives me a nice long season worth investing some labor in to establish the plants for a decent yield. But those of you farther south can still apply these techniques for a spring or fall or even winter crop of tomatoes depending on your location.

In the next issue, I'll report on a new trellising system from Holland that I loved for cucumbers and hated for tomatoes. In subsequent issues, I'll write about how to recognize whether your vining crops are vegetative or generative and what to do about it; analyzing yield data to boost yield; and other ways you can get more production from your valuable hoophouse space.

Andrew and Ann Mefferd own One Drop Farm in Cornville, Maine. Andrew also works as a trial technician for tomatoes at Johnny's Selected Seeds Research Farm.

This photo of hydroponic greenhouse tomatoes illustrates the lowering and leaning technique. When the vines get too tall, they are lowered and leaned to the left, wrapping around the end of the row. The same trellising system worked well for the author in his hoophouse. Notice how all the foliage below the first fruit cluster has been removed, a technique that helps improve air flow and increase fruit production at the top of the plant. Photo by Andrew Mefferd.

Increasing tomato yield

By Andrew Mefferd

In the previous article, I talked about a spacing change in my tomato hoophouse, which allowed me to plant 60 percent more plants than the previous year. Yet I was able to increase yield by 100 percent over the previous year. Now, I'm going to report on the improvements that led to the other 40 percent of my yield increase. Many of the principles can be applied to the field as well as the hoophouse, and crops besides tomatoes.

Learning from yield data

One of the biggest changes I made between the 2011 and 2012 growing seasons, besides increasing planting density, was to eliminate a few varieties that did not perform well for me in 2011. I got rid of the underperforming varieties and played with a tomato all-star team in 2012. The only way for you to figure out who your all-stars are is to do what any coach would do — keep statistics. Then make the time to analyze them once the season is over.

This doesn't have to be a time-consuming endeavor. What we do on One Drop Farm is to repurpose our springtime plant propagation house (a 12' by 20' double wall polycarbonate lean-to off our garage) into the summertime tomato packing house. We throw a white tarp over the structure so it is still light inside when the sun is out, but the heat doesn't build up as it does when it's in greenhouse mode.

A ventilation fan runs constantly during the daytime to keep air flowing over the tomatoes, to dry out any moisture and reduce post-harvest rot. As tomatoes are picked and brought into the tomato packing house, they are weighed in and noted on a pad that lives right next to the scale. At the end of the season, we put all the numbers into an Excel spreadsheet set to auto sum and we can see how each variety did for us.

The differences can be significant and surprising. In 2011, I grew a bunch of different heirlooms (all of them grafted), and some of them did much better than others. So I eliminated the under-performers, and landed on four varieties that seemed to work the best for me: German Johnson, Cherokee Purple, Striped German, and Valencia.

Last year, I noticed something about Valencia that I had seen the previous year, but worse this time. The plants got excessively vegetative early in the season. Essentially, they were putting all their energy into vines and leaves and not putting much into flowers and fruit. The vines were as big around as the fat part of my thumb, and the new growth was so curled and twisted, it looked like herbicide damage! At the end of the season, I could look back at my yield data and see that that excessively vegetative condition cost me earliness and yield.

Now, this excessively vegetative condition is a combination of the plant being young and naturally vegetative, and being grafted to a vigorous rootstock. This is significant for Valencia because none of the other three grafted heirlooms I was growing reacted as vegetatively. Being convinced of the benefits of grafting as I am, I'm not going to stop grafting, but I am going to drop Valencia and try to find another orange tomato that performs better UNDER MY CONDITIONS.

I emphasize the last part because this is why the yield data is so important. It's not that I think Valencia is a bad tomato. In fact, I think that for an orange tomato, it is remarkably uniform and unblemished, especially for an heirloom. But under my conditions it doesn't perform the way that I would like it to, and I would rather fill the space with something that can be more productive.

If I didn't graft or lived in another part of the country with different weather conditions it might perform differently for me. Getting the hot new variety or finding out what other growers in your area have success with may get you half the way to good production, but there's no guarantee unless you take the yield data and verify for yourself that what you are doing under your conditions is working. Knowing which varieties produce best is worth way more than the time that it takes to get the data.

Your yield data may really surprise you. I know it surprised me this past winter to find out that I was only getting one third the yield off my grafted heirlooms than off my grafted hybrids. I was shocked! All the plants looked vigorous and healthy. You would never think the yield difference was that pronounced just by looking at them. It is always said that heirlooms are not as high yielding, besides not having the disease resistance of hybrids. I would have guessed maybe a half or so of the hybrid production but a third!?! This is an example of something I never would have known if I didn't have the data.

So, what does this mean to me? The way I think about it, my grafted hybrid beefsteaks yield about 30 pounds per leader, or since they're double leader plants, 60 pounds per plant. I try never to take less than $2.50/pound for my red beefsteaks, and we sell a significant amount of them for more than that at the farmers market, so a grafted two leader hybrid beefsteak tomato plant is worth at least $150 gross to me.

The heirlooms, on the other hand are yielding closer to 10 pounds per leader or 20 pounds per plant. I didn't sell any heirlooms for less than $3/pound, and sold a significant amount for more at the farmers market, so I can say that a grafted two-leader heirloom plant is worth at least $60 dollars to me. But even that doesn't tell the whole story. It probably takes me twice as long to maintain a row of heirlooms as it does a row of hybrids. This isn't a coincidence; greenhouse hybrids are bred to be lower maintenance. The opposite of this are the heirlooms, which have more double heads, dead heads, excessively large flower clusters, and generally weird and unruly growth habits that make them interesting to work on but time consuming.

In case you're wondering about the raw data, I averaged 31.44 pounds per leader off my hybrid beefsteak plants, which were Rebelski grafted onto Colosus. On the heirlooms, which were all grafted onto Maxifort, I averaged 10.25 pounds per leader from German Johnson, 9.84 pounds per leader from Cherokee Purple, 9.7 pounds from Striped German, and 7.25 pounds from Valencia. Due to their excessively vegetative condition, the Valencias were also more than two weeks later to start bearing than the other heirlooms. And I only count marketable yield, since I leave the unsalable tomatoes in the hoophouse to be picked up with the other plant debris. Though the hoophouse does such a good job of protecting the fruit from the elements, I would say that I pack out somewhere in the high 90 percent range of the fruit that I grow.

And why did I use Maxifort for the heirlooms and Colosus for the Rebelski? I botched my first round of grafting with the heirlooms and by the time I started over again, Colosus was sold out. Most heirlooms seem to germinate and emerge in a much less even manner than most of the hybrids, taking a week or more for them all to come up, whereas the hybrids seem to all germinate over the space of a few days. When you need perfectly matching stems for grafting success, it can be advantageous to start four times the number of heirlooms than you really need to plant. The seed is cheap, and that should give you enough to match to your rootstock variety, which is more expensive and you could overplant by only 20% (or more, depending on how good you are at grafting).

Whether it is "worth it" or not to grow heirlooms is up to the individual grower. I plan to continue growing them, though on paper it seems that I should consider doing otherwise. If I didn't grow some I would miss them. Another consideration is that our top-quality early season heirlooms set us apart and give us something to offer to

A tomato plant before deleafing, at left, shows the three leaves below the fruit cluster, which is typical of indeterminates. Above, the same plant after the three leaves have been removed from the stem. Note how they have been cut off flush with the stem leaving no stub. Do this on the morning of a sunny day to give the wound all day to heal and dry out. Be careful not to cut the string! Photos by Andrew Mefferd.

suppliers that might otherwise not do business with us because we are selling what everyone else is selling. Another thing I keep in mind is that not everyone wants to buy a red tomato and we would surely lose some farmers market customers if we only had reds.

I figure as long as we can get 4,000 pounds of tomatoes at $2.50/pound minimum price out of a 30 by 48 structure that probably cost less than $10,000 to build, we are making the cost of the structure back every summer season. To my mind, that seems like a pretty good return, even though some varieties are more profitable than others. And knowing what different varieties are producing for you is golden when it comes to figuring out whether to continue with a variety, or even switch to another crop altogether.

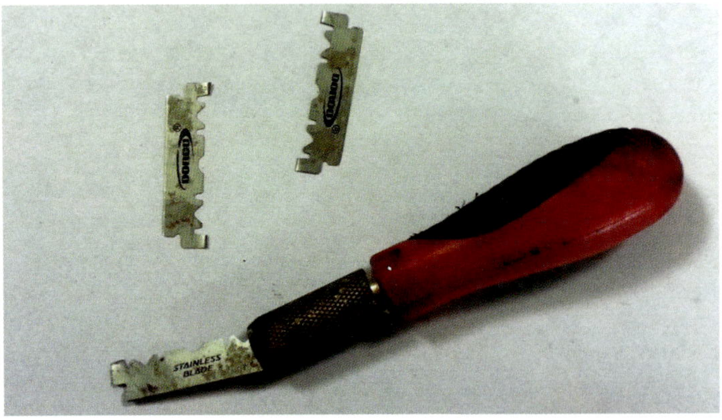

The author's deleafing tool: A razor blade snapped in half in a woodworking blade holder.

Leaf and cluster pruning

One other thing that probably helped out my yields last year was that I kept up on leaf and cluster pruning. By leaf pruning, I mean removing the leaves below the cluster of tomatoes that is ripening. This is important to do for several reasons. When the leaves are allowed to remain below the ripening cluster, it keeps the truss cooler and that makes the tomatoes ripen more slowly. So, taking them off can speed up the ripening process. Also, the cluster microclimate caused by leaf cover on all sides traps more moisture close to the ripening fruit. This increases the possibility of botrytis developing on the calyx or fruit, and russeting or micro-cracking, which is made worse by free moisture sitting on the surface of the fruit. I have found fruits that are in perfect condition except for some russeting where a leaf was touching the fruit and keeping the fruit wet. This is particularly a problem late in the season in an unheated hoophouse, when cold night temperatures cause a lot of condensation on everything including the fruit. The fall is really the only time of year that I cull fruit because of russeting.

The leaves below the fruit cluster are too shaded to contribute much to the plant, and as the oldest leaves they are the most prone to diseases. When you take them off it opens up a lot more air circulation at the bottom of the plants. Otherwise, this is an area that tends to stay cool and damp and build up diseases like botrytis that like those conditions. The best way to remove these leaves is to cut them off flush with the stem with an x-acto knife, or I have come to prefer an old-fashioned razor blade snapped in half lengthwise and inserted into a woodworking blade holder. Do not snap leaves off or in any manner leave a stub of petiole attached to the stem. Stubs are very prone to botrytis infection that can move into the stem and possibly kill the plant.

As far as cluster pruning goes, it helps the plant remain vigorous and keep fruit size up to thin the cluster back to the desired number of fruits. Every flower that becomes a fruit is essentially a growing point, not unlike a shoot, which makes a demand on the plant to put energy into developing the fruit. One of the ways that greenhouse hybrids are easier to maintain than heirlooms is that they produce a more reasonable number of blossoms than some heirlooms do. I have seen a large-fruited variety like Brandywine throw 15 or 20 blossoms in a cluster!

A variety like that with fruit size from a half pound to a pound is obviously not going to develop all those fruits and have a 10-20 pound truss! Help it out and cut it down to a more realistic number. Most people prune beefsteaks to 3-4 flowers, cluster tomatoes to 5-6 flowers, and cherry tomatoes very lightly or not at all. Cherries can be tricky because the number of fruits set varies so much by variety. But if you look at the plant and many of the clusters have undersized, undeveloped fruit that never ripen at the end, you could nip that number off the future clusters to save the plant from putting energy into fruits that are never going to develop.

This is also a way to influence fruit size. If you have a really large variety that you would like to be a little smaller, you could leave four flowers and you will get four, smaller fruit than if you had pruned to three. If you have a variety that is a little smaller than you would like it, you could prune to 3 fruits and the plant's energy for that truss will only be divided in 3, resulting in slightly larger fruit.

When to actually do the pruning is another question. If fruit is setting really well, it is common to do the pruning early on with flowers before the fruit is set. This keeps the plant from putting any energy into flowers and fruit that you don't want. On the other hand, sometimes fruit setting is poor and not every flower is turning into a fruit, due to excessively hot temperatures or inadequate pollination. In these instances it might be better to wait until the fruits have become the size of a small marble to thin, so you can see which ones will actually develop.

These last two techniques of leaf and cluster pruning probably require too much labor to make sense in the field. But they really do make a difference in the high-labor, high-yield environment of the hoophouse and greenhouse.

Andrew Mefferd and his wife, Ann, own One Drop Farm in Cornville, Maine. Andrew also is a trial technician for tomatoes at Johnny's Selected Seeds research farm.

Useful tools for the hoophouse

By Andrew Mefferd

I use a lot of tools and supplies in keeping up with my hoophouse crops. Some of them are very specialized and some are re-purposed. Here's a look in my tool bag.

Harvest tools

For harvesting I like having some of the curved grape and tomato shears on hand. We don't use them so much to reach around and harvest fruit while attached to the vine as advertised, but to clip the stems after they've been harvested. We leave the calyx on our large-fruited tomatoes, for nice presentation and to show that they're fresh. But sometimes we want to pick two layers deep into a black bulb crate, instead of putting them in a single layer in a flat. The stem where the fruit was attached to the vine would puncture the fruit on the top layer but we can use the curved shears to clip the stem off flush with the top of the fruit. Straight shears would leave a sharp edge on the stem from the angle you have to reach in and cut it off, so the curved ones offer a distinct advantage for this use.

Speaking of clippers, my new favorite tools are the line of Saboten harvest scissors. Hida Tool and Hardware Company features tools from Japan "that continue the metalworking traditions of the samurai sword makers." They carry three models of the Saboten harvest scissor, along with an interesting catalog of other farm, woodworking and kitchen tools. I was initially drawn to these clippers because of their very small size; I thought they would be good for harvesting and pruning in tight spaces inside the canopy of plants. They did prove to be great for this. Since they are smaller than your hand, they can go in tight spaces without disturbing what is around them.

The Saboten B-2 Harvest Scissors is worn like a ring, with the blade facing either in or out, freeing up hands when the scissors aren't needed.

The ergonomics are great, they are intuitive and feel very natural to use.

My favorite ended up being the model 1318 compact harvest scissors, because of the way they are designed to fit in the palm of your hand while using them, and they can be worn like a ring when you are doing other jobs with your hands. I particularly like this feature for jobs where I need to switch back and forth between using my hands and using a cutting tool. When trellising and pruning tomatoes, for example, I go back and forth between putting on clips, popping off suckers and cutting off foliage. It may not seem like a big deal but having to pick up and put down a tool hundreds of times in a day is one of those things that eats away at efficiency. Another nice feature is the safety, which keeps the scissors closed while using your hands, and is released with a quick flick of the thumb.

The high carbon steel blades are very sharp and make a cleaner cut than most clippers. The only downside that I have found is that the blades rust very quickly when left with plant juice on them in the warm, humid greenhouse environment. The blades are spring loaded so that when the safety is off, they pop open ready to cut. When they get rusty they don't pop open so well.

So like most things, if you want to keep them functioning well give them some TLC, wipe off the plant juice when done, oil them occasionally, and maybe don't leave them in the greenhouse. There is a very similar model to the 1318 called the B-2 which has an extra fine cutting blade tips, which don't make a big difference to me but might if you were cutting delicate flower stems or something else really small. There is another model 1315 that has a heavier blade than the 1318 or the B-2 but it's a little less compact and you can't work the safety with your thumb, you have to use your other hand to release, which I didn't like as much. So if you want to clip like a samurai, get some Saboten harvest scissors.

Trellising

Tomato trellis clips are more useful than simply for keeping tomatoes growing up a string, though they are great at that. There are several different designs that are basically a ¾" plastic ring with a hinge on one side and a latch on the other. They are sold in the open position, ready to snap on any vining crop that you want to grow up a string, like tomatoes, cucumbers, peppers, or eggplants, for example.

The latch keeps the clip securely attached to whatever you clip it to, and has a little tab on the latch so the clip can be opened back up after it is closed and repositioned or re-used. The hinge is large enough to accommodate a string the size of plastic or sisal baler twine. The interior of the hinge holds the clip where you want it to stay on the string, and keeps the plant from flopping over as it grows upwards. As the plant grows upwards, keep adding clips to keep it upright.

Clips come in several different colors of plastic and biodegradable options. I personally like the biodegradable option, because I am wary of reusing most supplies year-to-year in a greenhouse lest disease from one year survive to haunt you the next year. Even with sanitation there is a risk that some didn't get sanitized properly or that a pathogen survived on a piece of crud, and wouldn't that be a tragedy to see last year's diseases show up on your brand new crop?

Currently, the biodegradable clips cost more than the regular old plastic ones, but hopefully as the biodegradable ones become more common the price will come down. Another notable advantage of the biodegradable clips is that if you use biodegradable twine, the whole crop can be cut down, twine, clips and all at the end of the season and composted. It takes a surprising amount of time to take the clips back off at the end of the season. You may be able to save in labor the extra money you shell out up front for compostable clips.

I don't know about you but most farmers I know spend more on labor than supplies and this is one area where you may be able to save some labor money by spending a little more on supplies. Plus, in the hot, high UV greenhouse environment I've noticed that the plastic clips only last a season or two anyway, and may be prone to failure in successive years. Even if the plastic ones are a cent and a half each and the compostable ones are more like three, the karma you get from composting is priceless.

Left to right: Arch-type truss support; J-hook truss support; open compostable trellis clip from the top; open plastic trellis clip from the side; closed compostable trellis clip from the side.

We have developed a signaling system using these clips, so hoophouse workers can flag potential problems. When a hoophouse worker spots a potential problem with a plant, they put a clip on the string a foot or two above the plant with the problem, so there is a clip hanging in space not anchoring anything. Since there is no other reason to put a clip there, that has become our symbol for "assess this plant carefully before working on it," in my hoophouses.

Bright flagging tape would be even more eye-catching than a clip, but clips are what we are walking around with when we are doing plant maintenance so this doesn't slow us down to go get flagging tape. The most common reason we do this is if a vining plant has lost its head or has a damaged head, it is an indicator to the next person who works on the plant to leave a sucker or two from which to develop a new head. We may also put a clip above a plant that is showing signs of disease, as a signal not to work on that plant and then spread the disease to other plants, and make the plant easier to find and monitor. Or if I'm not around, my workers may flag a plant that has some miscellaneous condition that they don't know how to deal with, so later we can quickly identify the problem plants and figure out how to deal with them.

We don't always use them, but when we notice that our tomato plants have particularly weak stems on the fruit clusters, we sometimes use truss supports. There are two main forms of truss supports: the arched truss support and the J-hook. Arched truss supports look like a miniature plastic bridge that is snapped onto the truss where it comes off of the main vine. The arched type is usually used on smaller tomatoes from grapes and cherries up to cluster tomatoes. Because the clusters on these smaller types of tomatoes are never as heavy as beefsteaks, all that is needed to keep them from kinking is a little support to keep the stem in a proper arch.

Beefsteaks, on the other hand, can pull so hard on the stem that the most common way of supporting them is to attach them directly to the trellis string. This is done with J-hooks that have the appropriately named hook on one end and a clip for attaching to the trellis string on the other end. They are used by slipping the J-shaped end of the hook under the stem of the first or second fruit on the cluster, and clipping the other end of the hook directly to the trellis string. This is done when the fruit are small before they have begun to exert much downward force on the vine.This can be particularly important when the crop is young or any other time when the plants tend to be very vegetative and produce weak, spindly fruit stems. We have had heirloom beefsteaks get so large with the stems so weak that they have actually ripped off the plant!

Even if they don't rip off of the plant, a heavy load of tomatoes on a thin truss stem will most likely kink the stem. Some growers think this is a big problem and some don't, but if the stems are so thin in relation to the fruit load that there is any danger of ripping off, it might be worth putting on some truss supports—especially when you consider that the first few clusters are your earliest and likely most valuable fruits. If this problem occurs, it is usually just the first few clusters that are affected the most and once the plant settles down and becomes more balanced the stems will be better formed. For more on weak trusses and plant balancing, see the June/July GFM article on vegetative/generative steering.

Some people swear by them and some never use them but if you're having trouble with tomatoes damaging their own stems you might consider trying some truss supports. The other issue that can sometimes occur is that the kink in a fruit stem can provide an entry point for pathogens in greenhouses that are having disease problems.

J-hook type truss support in use. You can see how the one end hooks onto the trellis twine and the other end hooks onto the main stem of the truss (peduncle) after the first or second fruit, and suspends the fruit weight off the twine so the stem doesn't kink or rip. Once again, these don't need to be used unless this is occurring. Both of these truss supports need to be applied when the tomatoes are set but small and not exerting much force on the peduncle yet.

Ergonomics

For keeping all these small tools handy and easily accessible while working and staying hands-free, I like using carpenters nail pouches that you can get at the hardware store. Just tie them around your waist like a belt and they hold enough clips or whatever to keep you busy for a while. Since they are made of canvas, they can be laundered and sanitized as they get dirty or come in contact with pathogens.

It's not just because my Mom is a librarian that I like having a library kick stool in my hoophouse. For jobs that only require me to be a little taller, I really like the kick stool because I can push it down the aisle with my foot as I move along maintaining the crop and never have to stop what I'm doing to move it. The one I have is about 16" tall, made of plastic and very light. Spring-loaded wheels in the base push the whole structure up an inch or two off the ground when there is no weight on it. As soon as I put weight on it, the springs are compressed, the base makes contact with the ground and stabilizes the whole thing. The fact that the base is wider than the top which you stand on makes it more stable than, say, a bucket which I have also used with less satisfaction.

For taller jobs we have a couple of stepladders. The top wire in my hoophouses is at nine feet so lowering and leaning and other jobs that require us to work at the tippy top take a little more than the kick stool. It seems like most of the stepladders in the six foot range will get us up there and fit comfortably in our walkways which technically are four feet wide but more like two feet or less once the plants grow out into them.

For close to the ground work, we have come to appreciate the "garden buddy" type of planting carts that are designed to provide a mobile seat and toolbox in the garden. We have a few different ones, one metal with pneumatic tires and a tractor-style seat and some others that are more like boxes with wheels. In addition to sitting on them to make lowdown jobs like deleafing and harvesting less backbreaking, we have found we can bungee a bulb crate or other harvest container to the top of one and push it with our feet up the aisle as we harvest, so we don't have to pick up our bounty every time we move along the row.

A few basics

I always keep some duct tape handy, because in addition to everything else it can fix, it can patch up a tomato vine! In the past when we have accidentally snapped tomato vines when lowering and leaning or otherwise handling them, we have had almost 100% success at taping them back together with duct tape. I have heard that if the vine snaps completely off with no attachment left this won't work but every time we have broken them they have been at least hanging by a thread and this has always worked.

As quickly as possible after a tomato plant is snapped, line the vine back up as it was before it broke and wrap the stem below the break with tape. Wrap in a spiral up from the encircled area to contain the break entirely in duct tape and continue on a few inches above the break so the "cast" is supported on both side of the break by healthy vine and duct tape. It may help to put a J-hook or other small piece of clean material by the break to act as a stiffening "splint." Even on hot days when I was sure a broken stemmed plant was going to be a goner this has worked well and we have saved all the plants we have snapped. I'm sure it sets them back some but it's better than no plant at all. One time when I couldn't find the duct tape I used electrical tape successfully but duct tape definitely works better with its greater width and stickiness.

I always keep a spray bottle with some rubbing alcohol handy. As a certified organic grower, I figure my best offense against disease in the hoophouse is a good defense. Whenever I touch a plant part that is obviously diseased or a dirty area like the floor with my hands or a tool, I spray them off and feel confident that I'm not spreading the disease from plant to plant after that. This is especially important when doing a job like deleafing below the flower cluster, where you are at a higher than average chance of coming into contact with lower leaves that may be growing botrytis or other diseases. It doesn't do any good to cut out infected plant parts if you just move the infection on down the row! There is a disinfectant used in the livestock industry called Virkon which is effective against tomato mosaic virus and some other viruses if that is a concern, but I'm not sure whether it conflicts with organic certification to use it on tools. So as always, check with your certifier first! The greenhouse industry has pruning shears that are connected by a tube to a bottle of disinfectant, so that every time the pruners are squeezed disinfectant is sprayed on the blade. I have never needed these, I just try to keep a clean house but if you need them they are out there!

The last and most important tool I'd like to focus on is fingers, my tool of preference for any job that I can possibly do with my bare hands. It's usually faster to use the tools you were born with than anything else. For example I always try to snap suckers off with my hands rather than cutting them. When I can do it in the morning while the plants are turgid, suckers usually pop off cleanly and I don't have to fool with a knife.

I hope this gives you a few new ideas to try this year to improve efficiency and the quality of work done in your hoophouse.

Andrew Mefferd runs One Drop Farm in Cornville, Maine, with his wife, Ann, and does variety trialing for Johnny's Selected Seeds at the research farm in Albion, Maine.

Heirloom tomatoes much better in hoophouse

Heirloom tomatoes are in high demand by chefs and consumers, but they are notoriously difficult to grow in commercial field situations. Many growers, though, have found that they get excellent yields and quality by growing heirlooms in a hoophouse.

The hoophouse protects the tomatoes from many of the problems that commonly ruin heirlooms outside. Diseases caused by soil splashing on the plants are reduced when hoophouse tomatoes are grown with drip irrigation. Splitting, which is caused by irregular watering, can be reduced because the grower can provide a steady measure of water to the plants. Sunscald is reduced because of the lessened light intensity beneath poly.

Massachusetts farmers Jeremy Barker-Plotkin and Dan Franczyk evaluated 38 heirloom tomato varieties in 2001, both in the field and in the hoophouse. They found wide differences among varieties in the field, but the best results in all varieties came from heirlooms grown in the hoophouse. First harvest was 3 weeks earlier in the hoophouse than the field, the farmers wrote, and the early harvests were heavier than the earliest field harvests. Fruit quality was higher and cull rates lower, than in the field.

Hybrid tomatoes were grown simultaneously in the hoophouse; yields were higher than heirlooms, but not so much higher that the heirlooms were unprofitable.

"In our marketplace, unheated high tunnels do not produce the earliest tomatoes, so the colorful heirlooms help to differentiate our product from the earlier greenhouse tomatoes," the farmers wrote.

The three varieties that were found to perform poorly in the hoophouse were Cherokee Purple, Black from Tula and Tigerella. All others did great. The best performers in each color are listed below, with yield from field trials in parentheses after each variety, expressed as pounds per 100 row feet.

Black: Russian Black (104); Black Brandywine (99)

Yellow/Orange: Nebraska Wedding (174); Golden Sunray (134); Amana Orange (109); Goldie (105); Moonglow (100); Russian Persimmon (98)

Pink: Eva's Purple Ball (99); Brandywine (97); Crnkovic (96)

Yellow/Green: Garden Peach (111); Green Pineapple (107)

Striped: Gold Medal (186); Marizol Gold (173); Northern Lights (140); Striped German (130).

This research was funded by a SARE grant. For the full report, visit the SARE website at www.sare.org.

Greenhouse tips for hoophouse growers

By Andrew Mefferd

As I've visited many heated, hydroponic tomato greenhouses in recent years, I've noted several techniques that would benefit soil-grown tomatoes in unheated hoophouses. Many hoophouses are put up by growers who are more familiar with open field growing and may not employ the full range of management strategies available to make the most of protected cropping space. In this two-part series of articles, I will talk about four techniques that I think could be used profitably in an unheated hoophouse.

These techniques may involve increasing the amount of initial cost and labor that goes into a crop in order to multiply productivity. I know labor is always short on farms and many people may question why they would put more labor than they are already using into growing a crop. But protected growing space is precious real estate, and there are a lot of techniques that can help you get more out of yours. As long as you're getting more productivity or quality out of a GH/HH space than the value of the labor you're investing, you're coming out ahead. Investing $5 more in labor to make $10 more is a good investment.

A tray of grafted tomato plants. Photos by Andrew Mefferd.

Variety selection

It's winter time and you're looking at your seed catalogs. You notice there is a greenhouse tomato section in the seed catalog, and maybe your favorite varieties are not in it. Why not just go ahead and plant your favorite field varieties in your hoophouse or greenhouse?

Well, a lot of people do, which is why I make the point that growers could benefit from at least trying some dedicated greenhouse varieties if they haven't already. From visiting with growers I know that there are a lot of varieties (Big Beef, Early Girl, and New Girl come to mind) that are commonly grown in hoophouses though they were not bred for that. And, yes, they may do pretty well in a hoophouse. They may keep up with the yield of a greenhouse variety for a while. But over the course of a season a greenhouse variety will usually outperform a field variety.

One of the reasons is disease resistance. There are a bunch of diseases that are pretty rare in the field but common under cover, such as leaf mold and tomato powdery mildew, so even if the field variety is keeping up through the early part of the season, it's going to start falling behind if it comes down with one of these diseases.

The other reason GH varieties will usually do better in protected agriculture is because they have been bred and carefully selected to do just that. The GH varieties that get commercialized are picked because they have performed well under the conditions including higher temperatures and higher plant densities of the GH.

A good analogy for this would be horses. We have many different breeds of horses each bred for a specific task. If you had a thoroughbred, you wouldn't take it into the woods logging, would you? And likewise you wouldn't enter your draft horse into a race. Sometimes it's hard to see the difference without putting a few draft horses in to race against the thoroughbreds. So my advice is that if you've never tried dedicated greenhouse varieties, try a few plants or a row of one against your standard varieties this year.

I'm not saying you should not grow non-greenhouse varieties in your hoophouse. I usually grow about half a house of red greenhouse hybrids, and half a house of various heirlooms every year. I see the difference when I get a third of the yield out of the heirlooms than I do out of the greenhouse varieties. But that's a choice I'm making with eyes open, knowing what kind of yield I'm missing out on. I can count on the heirlooms getting leaf mold and stopping production late in the season when the greenhouse varieties are still chugging along.

Grafting

If you only try one technique from this article, I would say the most important one is grafting. Most of the big greenhouses are already grafting their tomatoes, but when I go to smaller farms it's hit or miss whether they're using grafted plants. Based on my experience on my own farm and doing variety trials on rootstocks at Johnny's Selected Seeds, I would say that grafting to a good rootstock variety like the ones that are currently on the market has the potential to add 30-50% to your tomato yields without compromising quality one bit.

It's worth noting that you can graft other greenhouse fruiting vining crops as well, like eggplant, cucumbers and peppers, but our results with those crops have been less clear cut. You can graft eggplant and peppers onto tomato rootstocks, and cucumbers onto other cucurbit species rootstocks. But developing good rootstock/scion combinations in those crops is trickier than in tomatoes, so that's why grafting those crops is not as common as with tomatoes. But it's definitely worth looking at and I'm sure we'll see more work on rootstocks for those crops in the years to come.

Grafting is the kind of technique where your mileage will vary depending on the conditions. It helps more where the conditions are less optimized...like in unheated, soil grown hoophouses! Larger hydroponic growers usually only get about 10 percent yield increase from grafting because they have already optimized most of the other growing conditions like temperature, fertility, humidity, etc. Most of the large greenhouses are grafting since that extra 10 percent is on top of an already very high yield, it is the icing on the cake. The increased profit on that extra 10 percent is still much more than the cost of using grafted plants.

Smaller growers generally have even more to gain from grafting. It is a generalization that obviously doesn't apply to everyone but many smaller HH and GH growers have not optimized production conditions to the extent that larger growers have. My own HH is a good example of this. Living in Maine as I do, every single night in my unheated hoophouses, even during the summer, is cooler than ideal, and many days during July and August the temperature is hotter than it should be. I grow in soil, so fertility for a long season crop is tricky, it's hard to tell how much of what nutrient is available at any given time, unlike in hydroponics where you know exactly what the plant is getting because they are getting a specific mix designed for the crop. Grafting is helping me overcome the production difficulties that I face to get a 30-50 percent yield increase from grafting. Let's break it down and look at why.

Interspecific hybrid vigor

I like to think of tomato rootstocks as the mules of the vegetable world. Mules are the cross of a horse and a donkey, two different species that produce offspring that has advantages over either of its parent species. One of the reasons mules can be better than their purebred parents is due to hybrid vigor. Hybrid vigor, whether in plants or in animals, is generally more pronounced when the two parents of the hybrid are more distantly related. My friend Donn Hewes says that mules have the toughness of a donkey in the size of a horse, and that's basically what a grafted tomato plant is. It has the toughness of a wild tomato with the culinary qualities of your favorite tomato, plus interspecific hybrid vigor.

For example, if you cross two tomato plants both of the same species Solanum lycopersicum, you may notice some increased vigor in the progeny from the hybrid. However, if you were to cross two tomatoes of different species, say S. lycopersicum and S. pimpinellifolium, you are likely to see a much greater boost in vigor than your cross of two S. lycopersicums. This type of cross of two different species is called an interspecific cross, between two different species. It is a natural way to get a huge vigor boost without turning to genetic modification or other more modern tricks.

Disease resistance

The technique of grafting is as much a plant breeding trick as it is a plant husbandry trick. What I mean by that is being able to make two plants into one frees the plant breeder's hand to work on the traits for each part of the plant independently from the other. Traditionally, one of the challenges for plant breeders was to give us a plant that has it all: high yields, good flavor, good disease resistance, etc. But it's hard to have everything we want because plant breeding is a balancing act and frequently adding to one area takes away from another one. For example, a breeder may start out with a great-tasting tomato that is hard to grow because it has no disease resistance. So the breeder crosses the good-tasting tomato with a disease-resistant one and most of the time, the flavor suffers in the process. This progression taken to the extreme can be found at your local grocery store, especially in winter. The grocery store tomato that we all know and loathe is an example of breeding that has focused on everything but flavor.

White-on-black plastic mulch is used between tomato rows for weed control and soil warming.

The beauty of grafting is that it frees the scion breeder to focus on the traits that would make a great tomato above the graft union (leaves and fruit) and the rootstock breeder to focus on what would make a great tomato below the graft union (mainly roots). The rootstock breeder can focus on just giving the plant a really great root system with lots of soilborne disease resistances, and forget about flavor, appearance, foliar disease resistance, etc. The scion breeder can forget about the root system and focus on fruit quality, foliar disease resistance, plant habit, etc. In fact, if you grow out most rootstock varieties they produce fruit that takes a long time to ripen, is not attractive and doesn't taste good. You would never want to use a rootstock ungrafted as a tomato plant but they do make great roots.

One of the common misconceptions about grafting is that the disease resistances from the rootstock extend up into the scion. This is not true. Each plant's own disease package applies only to its side of the graft union. For example, if you grafted a top variety that was susceptible to leaf mold onto a rootstock that was resistant to leaf mold, the top of the plant above the graft union would still be just as susceptible to leaf mold as if it had never been grafted.

That's why when you look at the resistance packages for rootstocks, they mostly stick to soilborne diseases. This is still a huge advantage, though, since it allows you to grow top varieties in soil you know to be infested with diseases that they cannot resist. The resistant rootstock acts as a blocker, keeping the disease in the soil from getting up into the top variety that is not resistant. That is why it is so important to plant grafted plants with the graft union above the soil line and keep the plants from sprawling on the ground, to prevent the scion variety from rooting in the ground.

If the plants are planted deeply, as is commonly done with tomato plants to develop adventitious roots off the stem, the top non-resistant variety will root into the infected ground and become infected itself. When grafting, it is advantageous to make the graft union as high up the stem of the rootstock as possible. If the graft union is just barely above the soil line, there is a good chance that the stem will sag and come into contact with the soil or soil will be thrown up against the stem causing the scion to root down into the soil.

Stress resistance

The last group of benefits that rootstocks can confer on plants is the ability to thrive in conditions outside the ideal, which cause stress on the plant. Because rootstocks incorporate the genetics of wild, tougher ancestors, they can make our cultivated varieties more resistant to stress.

Especially in unheated hoophouses, plants are often outside of the ideal high or low temperature, causing thermal stress on the plant. Rootstocks are selected for resilience in the face of thermal stress, and in this sense rootstocks can do some of the work in a HH that climate control would do in a heated GH. They broaden the temperature range in which plants will thrive.

Another type of stress that rootstocks are being selected to resist is water stress - drought, flooding and salinity. This means that plants will be less impacted by any one of these hardships, making the crop more forgiving for us to grow. Rootstocks are being developed specifically to deal with excess salinity, which could be useful wherever water is scarce and salts tend to build up, in places as diverse as Israel or Texas.

Rootstocks are also bred to resist chemical stresses, including pH being too high or low and nutrient stresses, too much or too little of essential nutrients. Wild species tend to invest more biomass into root production, so once again we see how grafting results in a superior plant where the rootstock does what it does best (root production) and the top variety does what it does best (fruit production). This superior root production includes both root density and rhizosphere exploration, which basically means that the root system is bigger and more aggressive about growing to seek out what it needs in the soil.

A bigger, more aggressive root system, in turn, can help the plant deal with both previously mentioned stresses. A bigger "pump" to draw from can help the plant through times when there might not be enough water, which may also help it through thermal stress by being able to pump more water. In times of excessive heat this may keep the plant from wilting simply by being able to keep up with the plant's water demand for longer.

I have seen the cumulative effects of all these advantages in my own hoophouses. In 2012, we had a very hot summer with many days in July and August in the high 90s and low 100s. I kept thinking that I was going to see some blossom drop, drop in production or fruit quality due to these excessively high temperatures. 2012 ended up being my best production year so far, and I think some of that success was due to the resilience imparted to my plants by their rootstock.

All of the tips in this article mean increasing up-front costs, be they seed cost of greenhouse varieties, propagation costs of starting more plants or grafting, or labor costs of lowering and leaning the plants. But any of these strategies that you may choose to do have the potential to increase the yield and profitability of the crop over the course of the season. Which is one of the ways that greenhouse growers achieve such high yields—by investing a bit more money and labor in the beginning to achieve an even greater return on the structure over the course of the season. You may not choose to do all of these things, but it is worth considering whether they may be able to help your 2014 crops.

Lowering and leaning

How do you grow a 30 foot tomato plant in a 9 foot space? Lowering and leaning! If you find your plants outgrowing the space they're in before the growing season is over, you might want to consider lowering and leaning.

Lowering and leaning is a trellising technique that was developed to be able to grow tomato plants in long-season greenhouses where the plant would outgrow the vertical space several times over. The basic idea is that instead of tying a string to a pipe or other solid object overhead, you attach a spool of greenhouse twine to a wire so that when the tomato outgrows its space you can lower it down a foot or two and give it some more space to grow. The leaning part of the two-step process comes in because after it is lowered, you move the spool down the wire so the plant's stem doesn't just coil up below the plant and possibly break. Leaning keeps the stem at a gentle angle to avoid breaking.

There are specially designed spools for this process, the most common being the tomahook but there are also rollerhooks and other devices that accomplish the same purpose. One other option is the Qlipr system I wrote about in the April 2013 GFM. Most of these devices require a wire instead of a pipe or other object overhead to hook onto, because of the size of the hook. But it may be possible to put very sturdy loops of wire everywhere a plant is supposed to be anchored in order to retrofit an existing pipe-trellis system.

Where I live in central Maine, the length of my growing season puts me right on the edge of needing to lower and lean. So I'd say for the people in the continental US who are farther south than I am (most of you), it's probably worth considering this technique. The top wire in my hoophouses is nine feet off the ground. In a year like 2012 when it was hot and the plants grew very fast, the plants grew up to the wire quickly. We ended up lowering and leaning many times and I think it was beneficial to have that extra space. In years like last year when we had cooler weather, overall poor growing conditions and the plants didn't grow as much it may not be worth it. I had installed tomahooks but never ended up lowering and leaning the plants because they grew so much less than the previous

Here's the view looking down on the author's tomatoes using tomahooks for lowering and leaning, dense spacing, white ground cover, and looking through insect netting that is used to keep out hornworms. Photo by Andrew Mefferd.

year, they were only just hitting the top wire when they needed to be topped at the end of the season.

This can be done with cucumbers and eggplant as well. Though you have to start leaning the eggplant to get it slanting in the direction you want it to go in before you ever lower it since the stem is so much more brittle than a tomato. Peppers can be trellised, but they are too brittle to lower and lean.

Planting density

Another possible yield increasing strategy is how many plants you put in a given space. I think planting density is worth thinking about because I go into so many hoophouses that have, say, five or six single rows across a 30' wide hoophouse. At this spacing the airflow is great but there could be a lot more plants in there! One way to look at it is if you go into your hoophouse and there is a lot of sunlight hitting the ground, that sunlight could be put to use growing plants. In my own hoophouse, I have emulated greenhouse spacing, with 5 double rows across the 30' house. I use a spacing for cukes and tomatoes with the double row two feet apart with four feet between double rows. And I've seen even tighter spacing used successfully. This has brought the amount of space used by each plant from nearly 7 square feet per plant in the first scenario to 3.5 square feet per plant in my current setup. The sheer number of plants in there has increased yield accordingly.

There is very little light hitting the floor in my hoophouse by the time the plants have hit the top wire. You might think the plants shading each other would reduce yield, but plants don't have an unlimited capacity for photosynthesis, and on a good sunny day, the plants will absorb as much light as possible even with a little shading. The thing I worried about the most when going to this heavy of a planting density was increased disease pressure with so much less airflow. And that is one possibility but with a good leaf pruning schedule to open up airflow down low and remove diseased leaves (see next month's article), I have kept my disease levels manageable and greatly increased yield.

I'd like to revisit the planting scheme that I mentioned in the March 2013 issue of GFM. That article discussed the use of a double row of tomatoes instead of single rows to greatly increase planting density. It is worth mentioning that instead of planting a double row of tomatoes in a bed, you can put the same number of plants in a single row. You then split the canopy into a double row by having two trellis wires or poles above the bed. The advantage of planting a single row at double density and training the plants to two overhead wires (making a double row out of a single row) is that weed barrier fabric can be laid right up to the single row of stems, leaving very little HH real estate uncovered, leading to very little weeding. This method has worked well for me, and is more similar to how hydroponic tomatoes are grown in a single channel and then trained to two overhead wires.

Some prefer to actually plant two rows to trellis up to two overhead wires. This allows for more air flow around the stems, and in hydroponic systems where people grow in containers like 5 gallon buckets, the buckets cannot be spaced closely enough together in a single row to accommodate all the stems. The disadvantage of planting two actual rows in the ground is that weeds can pop up in the space between the rows, unless you lay a skinny strip of weed barrier between the rows. And those weeds are hard to reach between the rows.

Andrew Mefferd co-owns One Drop Farm in Cornville, Maine, and is a senior trial technician at Johnny's Selected Seeds Research Farm.

Watering & fertilizing hoophouse tomatoes

By Andrew Mefferd

One of the best reasons to grow tomatoes in a hoophouse or greenhouse is to improve fruit quality. There are a lot of ways that fruit quality can be improved just by going under cover. You can keep off rain, which can cause checking or spots. You can keep the foliage dry, wet foliage being one of the necessary conditions for many of the worst tomato diseases (late blight, I'm looking in your direction). But controlling the amount of water the plant gets is an easy way to improve fruit quality.

One of the perennial banes of tomato growers is split fruit. It's easy to see after a few days' rain in a tomato field how the amount of water affects cracking. When it rains a lot and the soil becomes saturated, the tomato plant takes on a lot of water. If the fruit are ripe or close to ripe there is a good chance the skin will crack or the fruit will split open. A tomato's skin loses its elasticity when it turns red (or whatever the ripe color of the variety is), which is why cracking is the worst on tomatoes that are ripe or near ripe.

One thing you can do if you have problems with cracked fruit is to grow tomatoes under cover. But that doesn't automatically solve the problem. I've seen some hoophouses with pretty bad cracking problems as well. Because you now have the power to decide when it rains in your hoophouse, you have to make sure the way you irrigate supports your management goals. And when I say "make it rain" in the tomato house, I really mean "make the tape drip." I would never recommend sprinkler irrigation in a tomato house. Pretty much every foliar and fungal disease is made worse by moisture on the foliage and the spike in humidity that follows overhead watering.

The first thing about managing water to improve quality is to break irrigations up into smaller, more regular waterings rather than infrequent, big irrigations. In your greenhouse, you want to do the opposite of a big rainstorm dumping a lot of water all at once. The amount of water to apply on a given day varies widely depending on a lot of factors including soil type, temperature, humidity level, and many other factors. But let's say that on a given day you know your house needs an hour of water. Why not split that up into four 15-minute waterings? Or two half-hour waterings? If you are having problems with cracking, doing anything to divide the amount of water needed into smaller waterings should improve the situation.

I have had good luck using a DIG 9001D irrigation controller. Because I work away from the farm during the day, I find it very useful to pre-set the irrigation. The workers who are there are going to be busy with many other things and don't have time to fool around with irrigation. I know from experience and looking at the weather how much water my hoophouses are going to need on a given day. What I do is to use the irrigation controller to split the total amount of water I think the house is going to need on a given day (say, an hour) up into four smaller waterings (four 15-minute waterings in this case). Why four waterings? Well, the DIG 9001D can do a maximum of four separate waterings per day.

There are many other irrigation controllers out there, some of them with more options if you want to split your waterings up even more. I have found is that I can grow even crack-prone heirlooms and varieties like Sungold with almost no cracking at all. I start off with a lower amount of water early in the season when the plants are smaller and need less water and the temperatures are cooler, and increase the amount of water as the season progresses. If I know the day is going to be cool and overcast, I just suspend the irrigation program or reduce the number of irrigations.

Do you need more yield or do you need more flavor? The two are not always mutually exclusive but when it comes to watering, the management strategies for one are at odds with the other. For example, if you want more yield you may water more, to try to fill up more fruit faster. This principle can be applied to the field as well. When we grew more tomatoes outside, and in a hotter state (Pennsylvania), we would set up drip tape on the tomato block but turn it on only when the plants started to show the earliest signs of water stress, when the leaves would start to look limp.

This is a very "rule of thumb" kind of measure, but the principle is clear: less water tends to equal more flavor. More water tends to equal more yield. The year we were farming in Pennsylvania, we only turned on the irrigation once all season long, and our farmers market customers complimented the flavor of our tomatoes many times. Think of your tomato fruits as balloons. The amount of sugars and flavor compounds in the fruit are determined by the variety and the amount of sunlight the plant gets and how healthy it is. If there is more water in the tomato, the flavors will be more diluted. Since in a greenhouse you can control the amount of water the plant gets, you can prioritize yield by applying more water or prioritize flavor by applying less. The hydroponic equivalent of this is running your nutrient solution at a higher electrical conductivity (EC). A higher EC within the acceptable range will increase flavor and decrease yield.

Even the time of day you harvest can affect this balance. You can add a few percent to your yield totals by picking in the morning, before the plants have begun to respire and lose moisture, when the fruits are as full of water. I try to use the opposite strategy and pick in the afternoon when the fruits may not be quite as heavy but a little more flavorful. This also works for me because I

This is a Dosatron rigged up on a dolly so it can be used in multiple locations. The blue football-shaped object to the right attached to the manifold is a DIG9001D irrigation timer.

don't have a heated greenhouse, so my hoophouses tend to have a lot of condensation in the morning. Picking in the afternoon yields drier fruit that will have better shelf life, simply because it will not mold from the blossom or stem end as quickly as it would if packed wet.

Fertigation

One of the challenges to growing a crop over an extended season in a hoophouse or greenhouse is meeting the fertility needs over the entire time the crop is in the ground. Unlike a short crop like lettuce or greens, where there is the opportunity to renovate beds and apply fertilizer between relatively short cropping periods, long-term crops such as tomatoes, cucumbers, peppers and eggplant will be in the ground for a long time.

There is no way to pre-load the bed with enough fertility to meet the crop's needs for the entire season. If you applied all the nitrogen the crop needs at the beginning, the plants would be way too vegetative. And if you applied the full season's requirement of potassium at the beginning, it would tie up other nutrients. Even field tomato growers are recommended to fertigate, which means to apply fertilizer through irrigation water.

Once again we see that a big advantage of growing under cover is that you can control how much water gets to your plants. In the past few years, we have gotten into patterns of prolonged rainy springs here in Maine, and it has not been uncommon to see tomatoes in the early summer looking undernourished. We have seen spring rains wash the pre-plant fertility right out of the fields.

Even if you are growing under some type of protected culture, what the rain doesn't take the plant will. Over the course of a long season in a HH or GH, fruiting crops will deplete what fertility there is in the soil. This is one of the reasons that soil-grown protected culture crops will start to peter out toward the end of the season. Hydroponic culture solves this problem by delivering a small amount of fertilizer in every watering, evenly spoon-feeding the plant the nutrients it needs all season long.

Soil growers have the option to supplement fertility with either side dressing a solid fertilizer at the base of the crop, or injecting liquid fertilizer right into the drip lines. The advantage of side dressing is that you can use almost any fertilizer—compost, granular fertilizers or your own custom blend. The disadvantage of side dressing is that it is time consuming to go down each row distributing solid fertilizer. It can also be hard to apply it evenly.

Another problem with side dressing is that materials applied to the surface of the bed tend to mold in the high humidity, rainless environment of the hoophouse. Incorporating the fertilizer into the soil by scratching it in with a hoe can help. Or, if you use a plastic or biodegradable ground cover, pulling back the drip tape and ground cover, side dressing and then putting the drip tape and ground cover back over the fertilizer can help it stay moist under the ground cover and become available to the crop. But this takes even more time.

In the context of side dressing, the beauty of fertigation is that it takes almost no time. You just mix up your fertilizer solution and let the water take it away. The disadvantage of fertigation is that you are limited to a much smaller number of materials that are completely soluble in water and able to flow through drip tape. Always get assurance from the manufacturer that a product will not clog drip tape before using. The holes in drip tape are very small and easy to stop up with liquid fertilizer with a large particle size.

The piece of equipment that we and a lot of other farmers use with good success for fertigation is a Dosatron. What I like about the Dosatron is that it is calibrated so that if you know the concentration of your solution, you can apply an exact amount of fertilizer. Another useful thing you can do with a Dosatron is to put fertility into the water you are overhead watering transplants with. We use this a lot in the spring after our transplants have used up the nutrients in their potting soil to keep them green.

As equipment goes, the Dosatron is relatively simple, being a water motor that runs and dispenses fertilizer in response to the flow of water going through it. So there are no electrical parts to fail or short out in the wet, humid greenhouse environment. Another nice thing about the Dosatron is that you can either build it into the irrigation system of your greenhouse, or it can be put on a dolly so it can go from place to place if you need it in multiple locations.

Conventional growers have a lot more options when it comes to what will successfully flow through the drip lines for fertigation, since they can use any soluble fertilizer formulation available. Organic growers are limited to naturally occurring non-synthetics, many of which have a larger particle size than the chemical fertilizers. There has been a lot of work done on soluble organic fertilizers recently, and Hort Americas (www.hortamericas.com) is a good place to see what is currently available.

One of the trickiest questions about soil fertility in protected crops is how much to apply? There are established recommendations for most field crops, but it is hard to say how much protected crops with longer seasons need. Soil fertility can be hard to accurately assess since not all of the nutrients in the soil are immediately available.

The best way to tell what the plants need is to do tissue sampling on the crop, which is a service that many of the laboratories that do soil testing will also provide. You take some of the most recently mature leaves off your crop plants and send them into a lab and they will tell you what is actually in the plant's tissue. It is not that expensive and for anything larger than a very small hoophouse the service pays for itself to know that plants are getting low on a certain nutrient before you see a deficiency. You can actually see the nutrient levels in the plant go down over time as the soil becomes depleted. And that can be priceless for keeping yield and plant health strong late into the season when the plants and soil may otherwise be running out of steam.

Most prices I see are $15-35 per sample, depending on what is being analyzed, this is most useful if done weekly or bi-weekly. You can use it as a one-off test to diagnose a deficiency but it is even more useful to see how the nutrient levels change over time (i.e. generally go down over the course of the season in soil until growers start fertigating or sidedressing). So it adds up but priceless compared to lost production.

Some universities do tissue testing for farmers. A & L Labs, which has several offices to serve the whole country, is another option. You can find more information about tissue analysis on their website as well. (http://www.al-labs.com)

Andrew Mefferd, a frequent contributor to Growing for Market, is the co-owner with his wife, Ann, of One Drop Farm in Cornville, Maine. He also works as a senior trial technician at Johnny's Selected Seeds Research Farm. He can be reached at andrew@onedropfarm.com.

The proof is in the produce: two flats of crack-prone Cherokee Purple tomatoes grown without any cracks, thanks to careful watering. Photo by Andrew Mefferd.

Excluding insects from a hoophouse

By Andrew Mefferd

As we come to the end of a winter season that has resembled nuclear winter here in Maine, it is hard to imagine having to exclude bugs from our hoophouses. Yet the reality is that in just a few months we will have tomatoes and cucumbers planted in them, which is why we have our netting at the ready. I want to talk about our experiences using insect netting and other physical barriers to protect our crops.

In particular, we have problems with cucumber beetles early in the season in our plant raising nursery and hoophouse cucumber crop. What happens, at least in my area, is that cucumber beetles emerge hungry in the spring before field crops are typically planted. The result is that we experience very high pressure on our cucurbits in the hoophouse and seedling nursery. Our cucurbits under cover are the only thing on the menu when the cuke beetles come out but the field crops aren't transplanted yet!

The combination of high beetle pressure and the fact that we are certified organic means that we have limited options to deal with the legions of cuke beetles that descend on our plants. We have used spinosad or Pyganic to decent effect in the past. One of the good things about these products is that they break down quickly in the environment, which also means you have to spray frequently to get a significant amount of control under heavy pressure. Not only was this expensive in terms of buying the sprays and the time spent applying them, I would rather just keep the crop clean in the first place rather than having to spray it reactively. With the ever diversifying range of insect screens coming on the market, we decided to try screening two seasons ago.

Insect netting eliminated the need to spray organic pesticides in the hoophouse, which means the author's 2-year-old can eat cukes right off the vine. Photo by Andrew Mefferd.

We bought a roll of the 80 gram Proteknet insect screen from Dubois Agrinovation. It is a clear plastic mesh, which looks like it is made of very fine fishing line. It has openings that are 1 mm x .60mm (0.0394" x 0.0236"), small enough to keep cuke beetles out. We used it to cover the openings of the roll-up sides and the peak vents on our hoophouses.

We have double wiggle-wire track instead of a wooden hip board securing the plastic on the sides of our hoophouses. So we used the top channel on the double track to hold the plastic, and the bottom channel to secure the top of the netting. The sides of the net are secured in single wiggle wire track going down the ground posts, and the bottom is secured by stapling through used drip tape into the base boards. If you don't use double wiggle wire track to secure your plastic, and use a board instead, you could staple the insect netting at the top as well, UNDER the plastic of the rollup side.

Our tomatoes and cukes were planted in early May and we applied the insect netting shortly thereafter. By the time we were having some nice weather later in the month, the temperatures in the houses were spiking. On sunny days the temps were getting up to 100 degrees, and it was only May! We don't normally get temps that high in the houses that early in the season. The reduced airflow from having the netting on was slowing air circulation and making it much hotter in the houses.

Fearing temperatures that high would sterilize pollen in the tomato flowers and cause the blossoms to drop off without forming fruit, we pulled the bug netting off the side walls of the tomato house to encourage better cross ventilation. It was a good decision because 2012 ended up being a very hot one for us and we never suffered blossom drop or fruit quality problems. Even if the plants set fruit, temperatures that high can cause the tomato disorder known as graywall (also called blotchy ripening), which causes hard patches on the walls and/or core of a tomato that refuse to ripen. We left the netting on the peak vents and end walls for the tomatoes.

I left the insect netting on the cucumbers because as long as they have enough water, cukes can take higher temperatures than tomatoes. And they were the crop that really needed the netting to protect them from the beetles anyway. I wanted to protect the tomatoes from hornworms, which we'll get to in a minute.

But the cuke netting worked. Though it got very hot in there, the cukes were always healthy and there were a lot fewer cucumber beetles than in other hoophouses I saw in the area that were not protected. It's worth noting that there were still cuke beetles. They come in from somewhere, I'm not sure where, whether it be through cracks or when the door is open. But the netting kept them down to a dull roar which we found manageable, and worked well enough that we never sprayed our 2013 crop.

Temperatures continued to rise to unusually high levels in 2012, and we worried it was still too hot in both houses. When we built our two 30 by 48 houses, we left 8' wide spaces in the uprights at the ends of the houses, thinking that we could make a door and drive a tractor in if we ever needed to. We've never driven a tractor into those houses but I'm sure glad we left those openings.

We cut the plastic out of those openings on the ends, installed wiggle wire track around the openings, and put netting over the holes. This gave us a big 8' by 9' netted opening on the ends of our houses, improving ventilation even further but still excluding the pests. The one thing I worry about with this open-ended situation is violent thunderstorms. If one should blow up suddenly, it would be difficult to quickly secure the ends and strong winds could blow in and damage the crop or the house.

With the netting off the side walls of the tomato house, the plants were vulnerable to hornworms. But we have taken advantage of the obvious Achilles heel of the two species of moth that lay the eggs of the tobacco or tomato hornworm: they fly at night. We have been hornworm free for the last two years simply by rolling the sides down at night. If it weren't for the hornworms I otherwise wouldn't go to the trouble of rolling the sides down at night during the summer, but it works well to keep those pesky worms from pooping on my tomatoes.

Which still leaves the tomatoes vulnerable to one other even larger winged pest, er, farm resident. There is an ever-fluctuating number of loose chickens on our farm. Between my insistence that they not snack on the low hanging fruit, and our food safety plan, which dictates that they stay out of our crop production areas, we had to find a way to keep them out of the tomato house. So we took some old deer fencing that we weren't using with black plastic 1 ¾" mesh, and attached it where the bug netting had been. It keeps the chickens out and doesn't seem to interfere with airflow.

I am very happy with our use of physical barriers to exclude our hoophouse pests. I would definitely recommend it as a good way to prevent some pest problems before they even get to your crop. Make sure you are using the proper mesh size (Dubois has a number of different mesh sizes, and a guide online to help match mesh size to the pest you are trying to exclude). Just keep in mind that there is no way to exclude pests without also cutting down on some airflow.

Andrew Mefferd owns One Drop Farm in Cornville, Maine. He also works at Johnny's Selected Seeds, where is a senior trial technician. He can be reached at amefferd@johnnyseeds.com.

The Mefferds attached insect netting to the end of this 30' wide hoophouse using single wiggle wire track around the edges. The opening is 8' wide by 9' tall. The rest of the endwall is covered with poly, and the entrance door is to the left of the netted opening shown above. Two double rows of cukes are in the middle, with other crops on the edges. This piece of bug netting was sewn together with a sewing machine at the visible seam across the middle to make a piece large enough to cover the opening. Photo by Andrew Mefferd.

Pruning crops for health and yield

By Andrew Mefferd

Let's talk about all the great reasons to cut parts off of vining, fruiting hoophouse crops. In this article, we'll look at leaf and cluster pruning as applied to tomatoes, cucumbers, eggplant and peppers. Tomatoes tend to be the most management intensive of the four crops, so in most cases I will talk about the procedure on tomatoes and say how it varies for the other crops.

Most people who have grown tomatoes are familiar with pruning indeterminate tomato plants to a set number of vines, usually one or two, by snapping off the shoots or "suckers" that form at each leaf node. Whether in the field or greenhouse, sucker removal is a good opportunity to save some time and use your fingers to snap the suckers off. As long as the suckers aren't too big, say 1-2 inches long, they usually snap off neatly, and should not need special handling to heal properly. In fact, there are advantages to using your fingers to snap suckers off. Not using the same blade on each plant reduces the probability that plant diseases are transmitted from plant-to-plant contact. When I miss a sucker and it gets really big and won't snap off cleanly, I will use pruners or a blade to cut it off flush with the main vine. (More on the evils of "stubs" below.)

Disease prevention

There are a lot of great reasons to prune off parts of hoophouse crops that don't usually apply to field crops. In the case of leaf pruning, it is especially important in densely planted hoophouse crops to remove some leaves down low to promote airflow so the crop stays drier and healthier. And since the lower leaves are where diseases tend to start, it is a great preemptive move to take some lower leaves off before disease pressure gets high and spreads to other parts of the plant. If the lower leaves are not removed, they may be rotting on the plant by the end of a long hoophouse or greenhouse season.

Botrytis, otherwise known as gray mold, can be a season-long problem for hoophouse fruiting vegetable growers. The pathogen is endemic, there are no resistant varieties, and botrytis thrives in the type of humid environments that tend to develop in hoophouses and greenhouses. Temperatures below 65°F promote the development and growth of botrytis, so trying to save energy by turning down the thermostat makes the disease worse.

The area of the greenhouse likely to be the coolest and most humid is down at the base of the plants that are in the shade with dense leaf cover trapping moisture. Botrytis tends to start at the bottom and work its way up the plant through sporulation and by infected lesions contacting healthy tissue. Botrytis usually starts out on leaves but the real danger is when a lesion develops on a stem, because it can rot right through the stem and kill the plant.

If we had a heated greenhouse, we would try to minimize botrytis through climate control by keeping temperatures high and humidity low. Currently we only have unheated hoophouses on my farm, and we are able to manage botrytis pretty well with leaf removal.

When our developing tomatoes are close to their final size but still green, we remove all the leaves below the ripening cluster of tomatoes. By this time the plant is big enough to tolerate leaf removal.

Faster ripening

We harvest approximately one cluster per plant per week of large-fruited tomatoes during the summer season. Since there are three leaves between each flower cluster on indeterminate tomatoes, we aim to remove three leaves from each plant each week. As we pick the last fruit on any particular cluster we plan on removing the leaves above the finished cluster to expose the next cluster with ripening fruit. Leaf removal is a job we plan to do once a week.

One of the reasons I have heard not to do leaf removal is the rumor that the leaves on either side of a cluster of

Leaf and cluster pruning are two strategies that can help greenhouse and hoophouse tomato growers prevent disease and keep plants growing well. This leaf has botrytis; if not pruned off, the disease could work its way back into the plant and kill it

tomatoes are the ones that feed the cluster, but this is not true. The assimilates that are produced by photosynthesis are circulated through the entire plant. If you grew a two-headed tomato plant, and took all the mature leaves off one side, and all the fruit off the other side (so one half the plant was only fruit and the other half was only leaves), you would get roughly the yield of a single vine out of the two-headed plant. If the nutrients were not mobile within the plant, you would not get any fruit off the leafless vine. Plus, the leaves that you are removing from below the developing cluster are not performing much photosynthesis since they tend to be down in the shade and not getting much light.

The leaves on either side of a cluster of tomatoes create a microclimate, keeping the air around the tomatoes cooler and more humid. Tomatoes ripen faster the warmer they are. Humid air causes condensation on the surface of tomatoes which can lead to quality problems. So removing the leaves below the ripening cluster speeds up the process and keeps the fruit dry.

The other main reason I have heard not to do this is labor and that is always a balancing act. The grower must balance the importance of leaf removal against whatever else has to happen on the farm, but particularly for growers who have had problems with botrytis I would think this would be a priority.

How to remove leaves

It is important when removing leaves not to leave stubs. If you cut the leaf off but leave part of the petiole sticking out of the main vine, it will eventually die back and fall off. But it has a good chance of getting diseased and rotting back into the stem. When pruning you always want to cut back to a living plant part and not leave stubs of dead material protruding. I prefer to use an X-acto knife or old-fashioned razor blade snapped in half and stuck in a woodworking tool holder, but I know some people use bypass pruners. It is all about what is fast and cuts flush to the main stem. Do leaf removal in the morning or early afternoon of a day that is going to be sunny. Warm, dry conditions will dry the wound out and reduce the chance it will become infected. I carry a bottle of rubbing alcohol with me when doing any kind of pruning, and spray off my equipment and hands whenever I come into contact with diseased plant parts.

Taking off leaves also allows you to see into the crop better. The inner areas of the plant where airflow is poor and diseases would develop are otherwise shrouded by leaves. If you do find botrytis lesions on the stem of a plant, take a clean X-acto or other very sharp blade and cut the lesion out until all you can see is healthy tissue. Obviously it is important to catch the lesions when as small as possible, but a plant can function surprisingly well with a fairly large chunk of stem gone. There is usually enough vascular tissue left to feed the plant unless the lesion is very advanced. It is at least worth trying to cut a stem lesion out, since a botrytis stem lesion left unchecked will eventually kill the plant. It's also important when doing any type of pruning to clean up after yourself. Taking prunings out of the hoophouse at the end of the day and composting them away from your growing area is important. Dead plant tissue can easily become a source of the pathogens you are trying to prevent.

You can see how this improperly pruned leaf stub is slowly dying back to the main stem; in the meantime it is vulnerable to infection that could work its way back into the plant. It should have been cut off flush with the main stem

Leaf removal is not as important with cucumbers, peppers and eggplant as it is with tomatoes, because they are not as susceptible to botrytis. But if you are lowering and leaning tomatoes, cucumbers or eggplant, it will make it much easier to see what is going on with the vines as they are laid down at the base. And cucumbers in particular tend to re-grow suckers, which can be impossible to detect until they have gotten large if there are a lot of leaves obscuring the base of the plant.

Cluster pruning

Flower blossoms are another part you may want to consider removing. Tomatoes, in particular, will make more blossoms than they could possibly feed early in the season, or any other time when the plants are very vegetative. I have seen large beefsteak type heirlooms set huge clusters with 20 or more flowers. A tomato plant cannot possibly feed 20 beefsteaks on one cluster, besides the fact the cluster would tear from the plant under the weight of so many fruits. Every tomato flower is a claim on the plant's future energy. If too many fruits are on the plant, it will bog the plant down, divert too much energy from vegetative growth and make the plant weak.

Pruning the flower cluster back to a desired number of blossoms is also helpful in regulating the size of the fruit. The plant will only devote a certain amount of energy per fruit cluster. If more fruit are set on a cluster, the individual fruit tend to be smaller. If fewer fruit are on the cluster

This is the type of stuff you find when you prune off some leaves and look inside the canopy. The top fruit's calyx has been infected with botrytis which is rotting the fruit. Lower fruit is the characteristic boat shape—elongated flattened beefsteak shape with turned up ends like the prow of a boat, the result of two or more flowers fusing together early in the development process. The fact that it has 10 sepals is further evidence that blossoms fused.

This is an example of a flower cluster in desperate need of pruning. This is an heirloom beefsteak, I stopped counting the number of flowers at around 20. It could be cut back to the 3 or 4 flowers closest to the plant

they tend to be bigger. The cluster gets roughly the same amount of energy regardless of how many fruit are set.

Even though most of my tomatoes go to a farmers market customers and local wholesalers who aren't as particular about fruit size and uniformity as a grocery store, I still cluster prune. It helps to keep the plant balanced by regulating the amount of fruit on the plant at one time. And my customers who want beefsteaks don't want to buy the same variety in mini size, as can happen when too many fruits are set. They also don't want to buy the enormous tomatoes that grow when only one or two blossoms set on a cluster. There's something about at $10 tomato that is a psychological barrier for people.

So I pick a number of blossoms to leave on the plant, based on variety or type, and remove all the rest. I leave four flowers on a cluster for beefsteaks, five or six for tomatoes on the vine, six or eight for cocktail tomatoes, and don't normally prune grapes or cherries. But that is just a starting place—you always have to be aware of how large you want the fruit to be when making pruning decisions. Some people who want larger beefsteaks might leave only three fruits on a cluster. I might prune grape or cherry tomatoes if I noticed that every cluster was leaving five or six fruit undeveloped at the end of the cluster. I would nip five or six blossoms from the end of the truss to save the plant from wasting its energy on fruits that are never going to ripen.

I have a simple system that I use to quickly evaluate each cluster and move on to the next. First, I look for king fruit, which are the result of multiple blossoms fuesed into a single mega blossom. The first blossom, closest to the main vine, is much more likely than any other to be fused into a single blossom. The fruit that forms from this mega blossom is larger than normal and many times deformed fruit. These king fruit are the "boat shaped" fruit that some markets reject in favor of the normal flat round beefsteak shape. The main thing I don't like about them is they get huge enough to scare my customers away from buying them.

Another indication of king fruit is that their flowers are often jutting off at different angles from the other flowers on the cluster. If you look closely at a boat-shaped fruit you can visualize how it is multiple fruits that fused together at the flower stage. Another indication that a flower is actually multiple flowers fused together is the number of sepals. Most tomato varieties have five or six sepals, depending on variety. When you see a tomato flower with seven or ten or even more sepals you can be pretty sure that it is multiple flowers fused into one.

So the first thing I do is look at a cluster and see if the first flower closest to the plant is a king fruit. If so I cut it off. I look at the rest of the flowers on the cluster. I leave the desired number of flowers that are normal and closest to the plant, because the flowers closest to the plant fill out faster and tend to be better attached to the stem. In addition, the weight of the cluster is closer to the main stem so it's less likely to rip off or break.

If all your fruit are setting well consistently, you can do cluster pruning on the flowers before the fruit are even set. If you are having fruit set problems due to excessive heat or poor pollination, and blossoms are dropping before forming fruit, you might wait until the fruit are marble sized and clearly going to make it before deciding which ones to keep.

More or less the same principles apply to the other vining, fruiting crops. Small cucumbers, like cocktail and pickling cukes, don't need to be pruned at all. The fruit load should never get high enough to prevent fruit setting or disrupt the balance of the plant. Larger cucumbers, from slicer size to long European, should be pruned back to one fruit per node if more than one are set at a node. Likewise, peppers and full-size eggplant should have only one fruit per node, or both fruits may end up undersized. Or the plant may abort one of them after putting some energy into both. In fact, some pepper growers let a pepper set at every other node, depending on the condition of the crop. A general rule of thumb is that if the plants look weak on any of these crops, taking more fruit off than normal or even a whole cluster of tomatoes will cut the plant a break and let it put the energy back into the plant.

Why all this extra pruning? Well, hoophouse and greenhouse growing are all about taking your valuable, limited protected culture real estate and making the most of it. All of the practices described in this article are routine in the big heated greenhouses. They would not be done there if they did not make financial sense. There are a bunch of shrewd Dutch people who have already pushed the pencil on the cost/benefit ratio of all these practices. But maybe you've read this article and find yourself thinking "I'd never cut my tomato plant's leaves off, or my pepper plant's flowers off." And that's fine.

One of the things that I like about farming is that there are as many different ways to farm as there are farmers. At the end of the day you have to make sure your growing style suits you. Even if you don't use the techniques in this article, at least now you know they exist. You may find yourself in a situation one day where they are useful. I am a tomato geek and do all of these things because I like the overall management style and working on vining fruiting crops. I'm not trying to say this is the only way to do it. But if you find yourself suffering from botrytis low down or small fruit size, hopefully now you've got a few new tricks.

Andrew Mefferd is the co-owner with his wife, Ann, of One Drop Farm in Cornville, Maine. He also works as a senior trial technician at Johnny's Selected Seeds Research Farm. He can be reached at andrew@onedropfarm.com.

Using a tunnel for purposes other than growing

Many growers put their high tunnels to work even when the soil is fallow. These versatile structures can be used for all kinds of farm tasks. Here are a few examples:

- Hardening off transplants in spring.
- Drying onions and garlic.
- Curing sweet potatoes and winter squash.
- Hosting farm events such as dinners and workshops.
- Washing and packing produce.
- Storing tillers and other equipment on a temporary basis.
- Housing chickens and other animals (but be sure to observe food-safety and organic waiting periods for fresh manure.

Margaret Thorson of Thousand Flowers Farm in Waldron, Washington, said they use one of their three hoophouses each year for drying onions, wheat, beans and curing squash (shown above). "We rotate through them so we don't use the same one for storage each year. We have found that that can lead to mold issues," she said.

Understanding flowers can help you choose best varieties for hoophouses

By Lynn Byczynski

Any grower with a good market for cut flowers should easily be able to pay back the cost of a hoophouse in one to two years. Flowers perform exceptionally well under plastic, both by extending the season and by improving quality and yield.

Many varieties of cut flowers can be grown successfully in an unheated hoophouse. An even greater number are available to growers who heat their houses to a mimimum temperature of 45°F.

The chart on page 69 lists flowers that can be grown in either unheated or minimally heated hoophouses. But there are a few factors that you should take into consideration when deciding which varieties will do well on your farm.

Delphiniums can be grown in winter in a minimally heated house, or they can be overwintered in an unheated house for early spring bloom.

Day length

First, it's important to understand how plants respond to photoperiod or day length. Many flowers require a certain number of hours of daylight before they will bloom. Those that will bloom only in spring or fall are short-day plants, and those that bloom only in summer are long-day plants. Other flowers, which bloom whatever the day length, are referred to as day-length neutral plants. Still others are referred to as quantitative long day plants, which means they bloom fastest under long days but will eventually bloom regardless of photoperiod.

If a plant's bloom time is controlled by the day length, it is pointless to try to grow it in a different season in the hoophouse (unless you intend to use artificial lights to manipulate day length, which defeats the "low-input" aspect of hoophouse production). However, growing under plastic allows the grower to exploit day length response to the grower's benefit.

Take the example of sunflowers. Sunflowers will germinate and grow rapidly in cool temperatures, so they are good candidates for the hoophouse in spring and fall. However, photoperiod response varies by cultivar, so it's important to select daylength neutral cultivars such as Pro Cut, Sunbright Supreme or Sunrich. Those will bloom in 60 to 70 days in spring and fall at a much more manageable size than when grown in mid-summer.

Another example is Salvia leucantha, Mexican velvet sage. It is a short day plant that, when planted in early summer, produces vegetative growth during long days, then flowers when days get short in late September. For most growers, that means Salvia leucantha can't be grown in the field because a light fall frost will kill it before it produces enough flowers to make it worthwhile. But in the hoophouse it will be protected from light frosts long enough to get significant production from it.

Day length varies by latitude, so it's not possible to specify the dates that begin long days or short days. The longest day of the year is June 21, so the closer you are to that date (either before or after) the longer your days. In most of the continental U.S., the days are long enough to grow most flowers by the time it's warm enough to plant in an unheated house in March or April.

In a heated house, though, growers must be careful to grow day-neutral flowers or short-day flowers in the short days of winter. Long-day flowers cannot be grown in winter no matter how much the house is heated. It just won't work unless the grower also uses supplemental lights to "trick" the plants into believing the days are longer.

Although flower researchers always refer to "long

days" in regard to photoperiod, they now know that plants are actually responding to short nights. They have found that most "long-day" plants can be grown during the short days of winter simply by breaking up the long night into two smaller nights. This "night-break lighting" is accomplished by turning on lights for an hour or two in the middle of the night.

Temperature

The vast majority of flowers show temperature preferences. They may do well in the cool days of spring or fall, and limp along or even go dormant, in hot summers. Or they may languish in spring and fall and grow rapidly when days are hot in summer. Larkspur is an example of a cool-season flower and celosia is an example of a warm-season flower.

Other varieties may require cool temperatures during the seedling stage, but then be able to withstand extremely hot summer temperatures. Lisianthus is a good example.

Some flowers cannot be grown in the field or the unheated hoophouse, but may be excellent candidates for the minimally heated greenhouse because of their temperature and/or day length requirements. Freesias, for example, do not do well above 70°F. As a result, growers in most of the United States don't have enough time between the end of freezing weather and the onset of 70°-plus weather to get a crop. But because freesias do well in the short days of winter, growers can add enough heat to keep the tem-

Lisianthus, shown at left, is a valuable flower that grows best in the hoophouse, where it is protected from wind and rain damage. It also can tolerate high summer temperatures.

Sunflowers can be grown early and late in the hoophouse, as long as appropriate varieties are selected. Day-neutral varieties such as 'Sunbright Supreme', top photo, are single-stemmed hybrids that work well in spring and fall. Branching varieties such as 'Ring of Fire', above, are not a good choice for the high tunnel because they get too tall.

After light frosts have taken down all the field flowers in fall, hoophouse flowers can keep on producing until the weather gets really cold. Here, a crop of sunflowers and Salvia leucantha provide cash for weeks after the first frost.

perature at 45° and create the optimal conditions for this plant. Any flower that is day length neutral or a short day plant, and which needs cool temperatures, is a good candidate for a minimally heated house.

The most successful flower growers are those who can match day length requirements to their temperatures in the hoophouse. And this, unfortunately, is something that varies widely by location so it's impossible to provide an exact recipe for hoophouse flowers that will work on every farm. Each grower must learn the environmental requirements of each flower and then figure out what will work for him or her.

The best resource for learning day length and temperature requirements is *Specialty Cut Flowers* by Allan Armitage and Judy Laushman, available from *Growing for Market*.

Disease susceptibility

One last consideration in choosing flowers for hoophouse production is whether they are susceptible to foliar disease. Because of the reduced air circulation in a hoophouse, flowers that often get powdery mildew or other diseases are more likely to have problems. Zinnias, for example, are best suited for outdoor production because they are susceptible to foliar disease in the still air of the hoophouse.

Dutch iris require cool temperatures, so they can be grown in early spring in an unheated hoophouse or in winter in a heated hoophouse.

Recommended flowers for hoophouse production

Common name	Unheated hoophouse?	Heated (minimum temp.)	Comments
Anemone coronaria	yes	45°F	Hardy in Zone 7, anemones can be grown farther north in an unheated house. Plant in fall for early spring blooms. They also can be forced in winter in a minimally heated house.
Bupleurum, Nigella, Bells of Ireland, Agrostemms, Bachelor Buttons	yes	no	These hardy annuals can be direct seeded in fall in the unheated house to bloom two weeks before those grown outside. They are not high-dollar flowers but are geat fillers if you're making bouquets.
Delphinium	yes	45°F	Delphinium belladonna or hybrids can be planted as plugs in September and overwintered in a hoophouse for mid-spring blooms. Or they can be planted in late summer in a heated hoophouse for winter production.
Dianthus	yes	45°F	Sweet William types of Dianthus are biennials that should be planted as plugs in fall for early spring bloom. Several other tall varieties such as 'Amazon' and 'Purple Bouquet' do not require vernalization as the biennial types do, so they can be planted in late winter for bloom a few months later.
Dutch iris	yes	50°F	Plant bulbs in fall in unheated house to get early spring flowers. In a heated house, precooled bulbs can be planted all winter. They do not like temperatures above 65°F.
Freesia	no	50°F	Freesias take 120 days from planting to bloom but won't tolerate heat, so are not a good candidate for the unheated hoophouse. However, they do well in a minimally heated house in winter. After planting, the temperature should be 65° at night until five to seven leaves are visible; then it can be reduced to 50° at night.
Larkspur	yes	50°F	Plants require a cold period of 6 weeks under 55°F. Day length of 16 hours or more produces the longest stems. Heated production may not be profitable, but direct seeding in fall for early spring bloom will provide flowers two weeks ahead of those grown outside.
Lisianthus	yes	65°F	Plant plugs in early spring in an unheated hoophouse. They are slow growing, but tolerate cold and bloom in summer. After the first harvest, keep plants watered and they will rebloom.
Lupines	yes	45°F	Plant plugs in late winter for mid-spring bloom. Or plant in a heated hoophouse for winter to early spring blooms.
Ranunculus	no	45°F	Not a good candidate for the unheated house in areas where spring heats up fast. They take 90 days from planting of tubers and cannot tolerate temperatures over 60°F. But they're great in the minimally heated house in late winter because they like cool weather and short days.
Salvia leucantha	yes	35°F	Mexican velvet sage grows tall during long days but won't bloom till mid-September. If you don't usually have frost till October, plant in an unheated house. If frost is a threat in September, provide minimal heat, as it is very tender.
Snapdragons	yes	45°F	Snaps can be planted in fall and grown at 45°F night temperatures for bloom beginning in December. They also can be planted in late winter in the heated house for early spring blooms.
Stock	yes	45°F	Stock only does well at temps below 65° so it's marginal in spring when weather warms quickly. It's great in a minimally heated house in winter, though.
Sweet pea	yes	45°F	Sweet peas perform very well in the hoophouse, even in locations where they can't be grown outside. Start them early, six weeks before last frost in an unheated house or any time in a heated house. Provide a strong trellis because they are exceptionally productive.

Winter flowers are the most profitable

By Frank and Pamela Arnosky

POSITIONS WANTED - Two experienced cut flower producers are willing to relocate. Experienced in all aspects of production including but not limited to: floods, droughts, tornadoes, insect infestations, equipment breakdowns, sales, public relations, accounting, labor relations, personnel management, inventory control, fence building, computers, greenhouse and building construction, and much more. Able to navigate the complex workings of the IRS and the Immigration and Naturalization Service. Knowledgeable of EPA Worker Protection Standards, OSHA regulations, state regulations, and local regulations. Rapidly becoming bilingual. Able to work long hours in adverse conditions with little sleep. Minimum salary required per person--$125,000. Housing and beer must also be provided. Tropical or semi-tropical location preferred. Contact Growing for Market to submit position openings.

Bet you think we're joking?! Well, of course we are, but only halfway. Things are really looking bleak around here because of the drought. We might need to bolt! Since July 11, we have received only 1.25 inches of rain. The soil is bone dry several feet deep. San Antonio has had a little more rain than we have had, and even still, it is the driest year on record since 1951! Not the best conditions for cut flower production. Costa Rica would be much nicer!

So we are hunkering down, getting ready in case it gets really bad. We have been identifying our best profit makers to focus on, and production during the winter in cold frames is the clear winner. Nothing on the farm was as profitable, on a per square foot basis. Considering that the water usage is so much less, and we don't have to work in 100 degree days, this cold frame thing looks like the way to go.

We have one big "greenhouse" and six "cold frames". For us, the difference is that the greenhouse has benches, full heating and cooling systems, and is used for crops grown in specific climate regimes, such as poinsettias or plugs. Our cold frames are much less controlled. The sides raise for passive ventilation, some don't have heat, and the crops are grown in ground beds. They are really just a way to have better control on what would normally be a field situation. Our mantra for cold frames is "Keep it simple, stupid." The whole idea here is to keep overhead to a minimum. Our cold frames are 20 feet wide, and we can get four 4-foot wide beds in each house. The cold frames are 96 feet long, but our new one this year will be 140 feet long. The length is determined in part by the length of the plastic covering—the standard lengths are 100 ft. and 144 ft. Custom sizes can be made but they are expensive. The houses have a double layer of poly with inflation. In the houses with heaters, we keep the thermostat below 40 degrees. We have had good luck using floating row cover in our unheated houses. This can really give you some good protection under double poly.

We have two goals with our cold frames. One is to provide some income in the winter, and the other is to extend the season of some standard crops either earlier or later than normal. Our frames actually have something in them year round to take advantage of the bed space.

Winter production for us so far has been limited to crops that normally flower in cool, short day condition. In theory, it wouldn't be too hard to install night lighting to provide long days, but those houses would have to be segregated from the others to prevent light pollution from messing up the short day crops. In the meantime we'll stick to the easy crops.

Anemone and ranunculus are two stalwart crops for any cold frame production. In fact, ranunculus is the top money maker on this farm. But there are lots of other possibilities. Here are a few of them.

Sweet Peas

These are our favorite winter crop. They are a cash cow. Last year, we sold $2,200 of flowers from only two 80-foot long beds!! We got $4 for a 10-stem bunch — wholesale.

Sweet peas thrive in the cool temperatures of the hoophouse in early spring.

With the recent release of new winter flowering varieties, you can count on some reliable winter blooming. We use the Pan Am winter flowering series. They come in about 9 separate colors. The fragrance varies by color, but even those with the lightest fragrance are better than those that are shipped in.

We grow our sweet peas in single lines down the center on our cold frames, direct-sowing the seeds about 2-3 inches apart. Two runs of Tenax netting are stretched vertically to the top of the house. Sweet peas can grow 10 feet tall or more. Sweet peas can take a light freeze, but they will get burned below about 26 degrees, and 22 or less will kill them outright. Never let the temperature get below freezing after buds are forming. Constant mild temperatures will ensure the longest, highest quality stems. Constant moisture is also important since swings in soil moisture will cause bud drop.

Snapdragons

There are two types of snaps—field snaps and greenhouse snaps. For cold frame production, you'll want to stick with greenhouse types. Snaps are a tricky crop to grow, and not for the novice or weak at heart! They can be the prettiest thing you have ever grown, or a dismal, heart-wrenching failure. The greenhouse snaps are broken down into four groups that bloom according to season and day length. Group 1 is for winter production in cool, northern conditions, groups 2 and 3 are intermediate for longer days in spring and fall, and group 4 is for warm long day conditions. Using the wrong group at the wrong time is a recipe for failure. Catalogs that sell greenhouse snap seed will have a detailed calender of the groups and sow dates.

We start our snap plugs from seed, but many growers will buy in plugs. We plant six rows in a 4- foot bed with plants 6 inches in the row. Air space is critical, but so is bed space, so you want to grow them as close as possible. Greenhouse snaps are not usually pinched. Grow them on at 50-55°F degree minimum temps. They grow straight up and can get 4 or 5 feet tall. They can be truly spectacular. Snaps must be supported with at least one, but usually two, layers of netting. For local production, you can cut the stems when they are showing good color, but for shipping they have to be picked tight, with only one or two flowers open.

Snaps are susceptible to all kinds of plagues and vermin, which makes them so difficult. Rust is a major disease. Pythium root rot can be lethal, and botrytis and sclerotinia are problems in moist cool conditions. Spider mites have been our biggest pest problem. You won't no-

Two great crops for a minimally heated hoophouse are snapdragons, top left, and stock, left. Photos courtesy of Sunny Meadows Flower Farm, Columbus, Ohio.

tice them until you really have a case.

For all the trouble, locally grown greenhouse snaps can be a real money maker. We saw some beautiful snaps last fall in Boston that were fetching 9 to 12 dollars a bunch!

Stocks

These have been a real sleeper for us. We grow several beds of them each year, but we grew them late in the winter, and used a lot for our early bouquets. We really should be selling more of them as straight bunches. They are easy to grow, love cool temps and command a good price. We don't sell the flowers as much as we sell the fragrance!

Winter stocks come in from California at a pretty low price, and the quality is good. But one thing we have that they don't is the smell! For some reason, shipping stocks leaves a lot of the fragrance back in California. They can't compare to a fresh, locally grown stock for fragrance. We can get double the market price just because they smell so good!

Stocks can be grown just like snaps, only we space ours a little closer. We use 10 rows in a 4-foot bed, 6 inches down the row. Stocks can also be grown in an unheated house using row cover. They are fairly frost hardy this way, but the crop time will be increased. The main problem comes from growing stock too warm. The plants won't set bud in high temperatures, so keep that temperature down below 55.

Most stock varieties are only 50% double, but there are several varieties that are 95% double. These are the only ones to grow in cold frames, because growing space is expensive,. Seed is fairly expensive, but it comes up easily. We start all of ours in 392 plug trays and then transplant into the beds. Harvest the plants by pulling the whole plant up and cutting off the root. The only problem we have had to watch for is a late infestation of aphids on the buds. This is common, and you need to be on the lookout for them early.

Season extension

Lately we have been using our cold frames for season extension as much as for winter income. By this we mean that we are planting a summer crop a bit earlier in the spring than the field crop and getting a harvest several weeks earlier. Usually, we will clear out a winter crop of anemones or ranunculus in late February or early March, and have a batch of plugs ready to go right in the same beds. In really cold climates like Minnesota, this may be the most economical way to use cold frames. When the outside temperatures are below zero, it is still going to take an awful lot of heat to keep a house of anemones above freezing. By putting in an early crop of hardy annuals such as larkspur or lisianthus, you shouldn't have to heat a cold frame at all, and still get a good jump on your competition. Here are some things that we put in in late winter.

Lisianthus

The best crop of lisianthus that we ever grew was last year's crop that we put in on March 1st, right behind ranunculus. We had 392 cell plugs that went directly in the beds, spaced 6 rows per bed, 6 inches apart. The plants were tiny, and weed control was tough at first, but they grew into plants that were 3 feet tall with a huge spray of blooms. Three or four stems filled a sleeve, and we were getting 8 dollars a bunch. We were picking in early June, 3 weeks ahead of the field crop. We used no heat, and several times the seedlings were exposed to below freezing temperatures, without a problem.

Larkspur and delphiniums

We have done really well with both of these crops in a cold frame, but for us here the crop time was a bit too long for us to plant after our winter crop. They needed to be planted in late December, and that essentially took up a bed for the whole winter season, just to get a two-week advantage on the field crop. But if you have an unheated house in the North where you are not using your beds for winter production, these would be great crops. They don't need to be heated, and should survive even in the coldest climates with a little bit of row cover.

We used the standard 'Giant Imperial Strain' larkspur, and 'Pacific Giants' delphiniums. The delphiniums were too tall, and we'd recommend one of the dwarf varieties such as 'Clear Springs' or 'Magic Fountains' for cold frame production. They will be more manageable. Watch for powdery mildew and sclerotinia with both these crops.

Celosia

The big crop that we use our late season cold frames for is cockscomb celosia. As all the winter crops finish up, we go right back in with celosia plugs. We can get a jump of well over a month over our field crop, with no chance of them bolting from cold soil. These celosia are one of the best moneymakers for our cold frames, and they dovetail right into any crop that is finishing up. When most folks are just getting their plants out, we are selling top quality stems! Plant them just like lisianthus--6 rows by 6 inches, and use netting.

Pamela and Frank Arnosky are the owners of Texas Specialty Cut Flowers in Blanco, Texas. They wrote about flowers for many years in Growing for Market. Many of their articles are collected in the book Local Color, available from Growing for Market.

Managing soils, weeds, pests, diseases

By Lynn Byczynski

The benefits of growing produce and flowers in hoophouses are well-documented: higher yields, better quality, a longer season, even better flavor in some crops. But hoophouses come with their own set of problems. Among the prominent management challenges are soil quality, fertility, insects, diseases, and weeds.

So far, research on these topics is limited to just a few projects around the country, and this article will summarize the most relevant.

Salt buildup

Many growers worry about the soil in their high tunnels. Outside, soil texture changes with the seasons in response to rainfall, freezing and thawing. Inside the hoophouse, though, the soil rarely freezes and is not exposed to rainfall at all. It seems intuitive that soil biology and structure must be affected by being continuously covered by a hoophouse.

High tunnel growers should be aware that keeping the plastic on tunnels year-round can result in a buildup of salt in the soil that can damage some salt-sensitive crops. Seedlings are especially susceptible to salt.

Salt-sensitive plants are less able to take up water from saline soils and can become water stressed. Sodium, chloride, or sulfate levels may become toxic and cause nutrient imbalances with potassium and calcium. Yields can be reduced significantly. Other symptoms of salt injury include necrosis (burning) of the leaf margins, stunting, wilting, and even plant death.

According to Dr. Elsa Sanchez, associate professor of horticultural systems management at Penn State University, the first step to preventing crop damage from salt buildup is to do a soil test every year. The soluble salts test may be an addition to the basic soil test, with an additional fee, so be sure to request it when you take your soil sample to Extension. When you get the soil test results, compare it to the chart on the next page to determine if you have a problem that needs to be addressed.

Soluble salts accumulate because the high tunnel covering prevents leaching by rain and snow. Also, most tunnels are irrigated by drip irrigation, which also limits leaching. Both organic and inorganic fertilizers can cause elevated levels of soluble salts. Compost, especially, can raise soluble salt levels quickly.

In a two-year nutrient management study of Penn State's high tunnels, applying even 1 inch of compost provided excessive amounts of nutrients. The baseline soluble soil level tripled in one year and increased six times in two years — to a level that could reduce yields of carrots, onions, bell peppers, lettuce, sweet potatoes, broccoli and potatoes. When 2 inches of compost were applied, soluble salt levels increased 13 times in two years, to levels considered strongly saline.

Plant response to soil salinity levels

Salinity level (mmhos/cm)	Effects
Less than 0.40	Negligible salinity; salt-sensitive cultivars of beans and carrots may show effects.
0.40—0.80	Very slightly saline; 25-50% decrease in yields of carrots, onions, peppers, lettuce
0.81—1.20	Moderately saline; seedling injury possible; 25-50% decrease in yields of broccoli, potatoes, plus plants above.
1.21—1.60	Saline; beets tolerant
1.61—3.0	Strongly saline.
Greater than 3.2	Very strongly saline.

Source: Pennsylvania State University, Agricultural Analytical Services

Dr. Sanchez suggests these ideas for preventing or dealing with high levels of soluble salts:

- Build new high tunnels in areas with good drainage to promote leaching.
- Avoid overapplication of nutrients. Using soil test reports, apply only the amount of nutrients plants need.
- Select fertilizers with low salt indexes. Avoid the use of nutrient sources containing animal manures, which tend to be high in salts.
- Use irrigation water with low salt levels.
- Use sprinkler irrigation to establish seedlings. Sprinkling can help leach salts around the plants.
- Leach out salts before planting. As a general guideline, 6 inches of water will leach about 50 percent of the salts, 12 inches will leach 80% and 24 inches will leach 90 percent. Sprinkler or drip irrigation can be used to leach salts. Drip irrigation will have to run for many hours to reduce soluble salt levels significantly. Penn State's Commercial Vegetable Production Guide (see below) specifies the number of hours needed to apply 1 inch of water, which depends on row spacing and the drip tube flow rate. For example, with a 0.45 gallons per minute flow rate, and rows spaced 4 feet apart, it takes 9 hours to apply 1 inch of water.
- Remove the high tunnel coverings and leave them off as long as possible so the soil is exposed to rain and snow.

"Out at the high tunnel facility, I've been managing four high tunnels since 2003," Dr. Sanchez said. "In the fall of 2007, the soluble salt level was on average 0.40 mmho/cm. That November, the tops of the tunnels ripped off due to high speed winds and because the plastic was getting old. We decided to leave the tops off until the spring. In April 2008, we put new tops on and had the soluble salt level of the soils analyzed. On average, the soluble salt level decreased to 0.09 mmho/cm or about 77 percent. Between November of 2007 and April of 2008, we got about 11.5 inches of rain."

Soil structure

Beginning in 2005, researchers at Kansas State University surveyed 81 growers with a total of 185 high tunnels in Missouri, Kansas, Nebraska and Iowa. Fifty-four percent had no concerns about their soil quality, but 14 percent felt there were problems and the rest were uncertain. Hardpans were reported by 32 percent and mineral surface deposits by 30 percent.

Sharon Knewtson, who was at that time a Ph.D. student at Kansas State, visited 79 of those farms and took soil samples inside the high tunnels and in adjacent fields outside. She tested the soil for multiple soil quality indicators including pH, salinity, particulate organic matter carbon and water-stable aggregates.

Her conclusion: Soil quality as measured was not negatively affected by high tunnel structures over time. Management practices such as tillage depth or frequency, application of fertilizers, use of cover crops and use of fallow periods had no effect on soil quality. Nor was there a correlation between soil quality and the length of time a tunnel had been in use, with all the tunnels at least three years old and some as old as 10 years.

While the overall message from that study is to relax about soil quality, two recommendations did emerge. The first is that most hardpan problems were caused by construction of the hoophouse and by restricted traffic patterns, e.g. paths. Growers should be aware of the possibility of a hardpan and if one develops, correct it with a deep-tined ripper. The second issue concerns mineral deposits (a white crust) on the soil surface, which may indicate a salt accumulation. Knewtson measured soil salinity in 93 tunnels at 63 farms and found that it was "slight and mostly superficial."

"The surface mineral deposit reported by 30 % of the survey respondents was not a cause for alarm," she wrote. "Electrical conductivity was not different in high tunnels with and without reported visible surface minerals. The mineral deposition at the surface could be carbonates, or salts, but as both pH and salinity were within acceptable limits at nearly all locations the presence of a surface mineral deposit may not be of concern."

However, because even slightly elevated salt levels can cause a 10 percent yield decrease in tomatoes and lettuce, she noted that "most growers can avoid yield reduction by leaching salts deeper in the soil profile with heavy irrigation before planting."

Fertility

Because most plants grow so rapidly and exuberantly in the protected environment of a hoophouse, growers need to pay particular attention to soil fertility. A soil test at the start of the growing season will provide recommendations about adjusting N-P-K and soil pH. But most experienced high tunnel growers also provide regular infusions of organic matter and fertilizer. John Biernbaum, a high tunnel expert at Michigan State University, says that too much nitrogen causes problems in high tunnels, so he avoids manure and instead uses compost made of plant materials. At the MSU student farm, compost is applied before every crop — two or more times a year — at a rate of 25 gallons per 100 square feet (a 5-gallon bucket every 20 square feet). Eliot Coleman, author of *Winter Harvest Handbook,* makes a plant and seaweed compost that he applies at a rate of 41 gallons per 100 square feet. And Steve Moore, another veteran high tunnel grower at North Carolina State University at Goldsboro, uses 12-27 gallons per 100 square feet.

Other organic fertilizers can be used instead of compost. For four years, the MSU student farm used a Bradfield alfalfa-based 3-1-5 fertilizer instead of compost. The recommended rate of that product is a half gallon per 100 square feet. Plants grew just as well as in the composted beds, Biernbaum said.

He advises that high tunnel beds should be broadforked (see Johnny's Selected Seeds or Valley Oak Tools for broadforks) to loosen and aerate soil without inverting it. Then they should be raked smooth and, finally, the compost should be spread. Irrigation, plant roots, and earthworms carry nutrients deeper into the soil over time, and having the compost on the surface at the beginning of the crop improves water absorption and seed germination.

Weeds

Weeds grow as vigorously as crops in the favorable conditions of a high tunnel, so weed management is even more important inside than outside. The oldest research recommendations were to cover the entire surface of the high tunnel soil with black plastic mulch, and many growers still do that. Plastic mulch has management problems, too, such as how to get the mulch tight on the soil surface and plant through it when the size of the tunnel doesn't allow for mechanical mulch laying. Another issue is repairing drip tape underneath plastic mulch if it gets damaged by rodents. And in some locations, black plastic mulch inside a hoophouse heats the soil too much.

For some growers, those problems plus the resource

and disposal issues make plastic mulch a non-option. They prefer to grow in bare soil and make weed management a priority. Here's what Biernbaum teaches at MSU about high tunnels and weeds:

"When starting with a clean greenhouse, the strategies are similar to working in the garden. At the top of the list is not letting current weeds go to seed. It is also important to observe what weed seeds may be blowing in from out side. At each opportunity, fallow ground should be watered and maintained to allow weeds to germinate, and then quickly and shallowly cultivated to clean up the space. Since weeds will typically only germinate from the top two inches of soil, cultivation methods that avoid exposing the seed in deeper soil will usually reduce the amount of seed germinating."

Carol Miles at Washington State University received USDA funding in October for a project studying biodegradable mulches in high tunnels in Washington, Texas, and Tennessee. Miles has previously researched biodegradable mulches in the field. Reports from that research are available at http://vegetables.wsu.edu/plasticulture.html, and future results of the high tunnel work will be posted there as well.

Insects and diseases

In general, crops grown in a hoophouse are less likely to be damaged by insect pests or diseases than those grown outside in the field. However, the same pests found in heated greenhouses — aphids, whiteflies, and mites — do sometimes create problems in the hoophouse. As Terrance Nennich of the University of Minnesota writes, "Remember that in the ideal environment of high tunnels, insects can multiply at very fast rates, and waiting too long for control measures could cost you the crop."

At Michigan State, the winter cutworm *Noctua pronuba,* which is much more cold tolerant than most cutworm species, destroyed a high tunnel of cool-season crops. The problem was solved by putting chickens in the tunnel for a month. Aphids have also been a problem on occasion.

Temperature is a key factor influencing the development of insect pests," Biernbaum writes "With freezing temperatures during the winter, there are essentially no insect pests. As temperatures increase in February and March, while the greenhouse is still mostly closed and there are no natural predators, aphid populations can begin to develop. The warmer the temperature, the more likely a population will grow to damaging proportions. It is true for most all greenhouse insect pests that temperatures above 60-65°F will lead to much faster insect growth.

"The other factor for aphids appears to be soil nitrogen availability. With increasing nitrogen availability or, more correctly, nitrogen excess, the probability of seeing aphid populations increase to damaging proportions is much greater. We have learned not to use excess compost or organic nitrogen fertilizer, particularly as the soil is warming and biological nitrogen is naturally increasing."

Tomatoes, the most popular crop for high tunnels, seem more susceptible to some typical problems inside than outside. Hornworms can multiply rapidly and do serious economic damage if not controlled. Closing the high tunnel in the evening will exclude the sphinx moth, which lays eggs that grow into hormworms.

Another increasingly serious problem of high tunnel tomatoes and other crops is white mold, Sclerotinia rot. The fungus that causes the disease, Sclerotinia sclerotiorum, overwinters in the soil, and in spring the fungal fruiting bodies emerge and spores are released. They will be released continually throughout the spring and can be carried on the wind into high tunnels. According to the University of Delaware Cooperative Extension, the biocontrol Contans has been effective in managing Sclerotinia diseases in the field. It can be sprayed in the area around the high tunnel and watered into the soil to reduce spore formation in the future. The spores can be carried less than 330 feet, so that's the distance from the tunnel where sanitation is important.

Resources

"Dealing with high soluble salt levels in high tunnels" by Elsa Sanchez: www.extension.psu.edu/vegetable-fruit

Penn State 2011 Commercial Vegetable Production Guide: http://extension.psu.edu/vegetable-fruit/production-guides/2011-comercial-vegetable-guide/pennsylvania-comercial-vegetable-guide-2011

Part 2 Building your hoophouse

By Dan Nagengast

It is important to think through your needs and understand the purpose and capabilities of a high tunnel or hoophouse before you begin construction. Those decisions will impact where and how you site the tunnel.

Keep in mind that the reason you are building a tunnel is because of its high potential production and the intensity of use it affords. A high tunnel will make your best ground much more profitable. The ideal site is on productive, level ground. Most houses will benefit from wind protection, if it is far enough away to prevent shading early or late in the season. The orientation (East-West or North- South) depends on your location; see the explanation in the next section.

You absolutely need a handy source of water, preferably one suited to drip irrigation. While watering needs may be less during the winter, you may still need to occasionally water, and hundreds of feet of frozen hose on a wintry morning is a nightmare. A few owners may also find they need electricity for certain crops. Though high tunnels are meant to be low-cost, high-return structures, there are some very profitable crops that need supplemental lighting in the winter. You may be considering some form of supplemental heat requiring electricity for harsh weather also. These considerations, and the need to visit and work in the house often, mean a site near the farmstead is most desirable. The bonus is that springtime in winter can be found just a few steps from your door.

Site preparation

Ideally, you begin with good soils and level ground. If necessary it is best to amend and till the soil before you begin construction. Level the area, even if it means hiring a grading contractor. Your house will be most productive if it is built on a slightly elevated pad, at least 3 feet larger than the house in all directions. While the contractor is there, think about whether you will be building additional hoophouses in the future and go ahead and prepare a pad large enough.

If you need to run water lines, do them once you have graded the pad. You might use the same trench to run electrical wire if you feel you will need it. Some may choose to place a hydrant in each house. We have had few problems with one hydrant located outside, but accessible for 4 houses. We keep the rolled up empty hose stored in one of the tunnels, where it becomes pliable quickly as soon as sunlight reaches the house in the morning. Make sure that your water line does not run where you will soon be driving ground stakes to erect the tunnel.

Laying out the tunnel

Laying out the footprint of the high tunnel is the most critical step. Typically, columns or ground stakes are driven 24 inches below grade down the long sides of the tunnel. The tunnel must be square and these stakes precisely located in order for the rest of the installation to proceed without problems. Failure to place the ground stakes correctly means the difference between easy construction and a struggle, and may even affect the integrity of the finished tunnel. So spend time doing this step, and don't be afraid to start over till you get it right.

You will be using string to lay out the footprint of your tunnel at exactly the height of the tops of the ground stakes above grade. Then the ground stakes will be driven in at the correct spacing down each side the length of the tunnel. Ground stakes are generally made of the same pipe as the hoops, but with a swedged, or compressed, end so that the hoops fit over them in a female/male arrangement.

Eventually, the hoops will be assembled and erected on the stakes, then connected by rods called purlins that run the length of the house. Hip boards will be attached along the sides to hold the plastic cover and accommodate roll-up sides. Base boards will be run at ground level, and attached to the outside of the hoops. The final construction step is to build the end walls and run aluminum extrusions to attach the plastic along the hip boards and over the hoops at either end. All that remains is to attach the plastic covering the top and sides, put on the end walls, and install a method to roll up the sides.

You can use a transit level to lay out the footprint of the tunnel, but it is also possible to use masonry string, batten boards, marking flags, a 100-foot tape measure and a string bubble level. Masonry string is available at hardware stores. It is highly visible dayglo in color, stretchy and strong. The string will be stretched across batten boards which are commonly used to lay out buildings. You will need eight batten boards, two for each corner. They consist of two picket legs to be driven into the ground with a board between them. Make your own using stakes and 1 x 4's., as illustrated below.

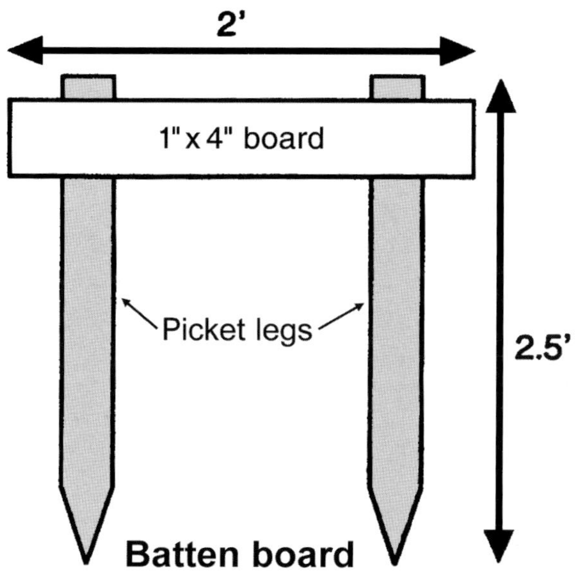

Establish one end of the greenhouse using a tape measure. For example, on a 20' x 96' tunnel, you would lay out a 20' width and mark it with flags or stakes. You now need to establish right angles at both ends to lay out the side columns. Here is where you can actually use the Pythagorean Theorem to determine plot squareness. As you recall, the theorem states that the square of the hypotenuse is equal to the sum of the squares of the other two sides on a right triangle. This is fairly straightforward on a 3-4-5 right triangle, or for that matter, any multiple: 6-8-10; 30-40-50, etc. All will have a 90 degree square corner, as illustrated below.

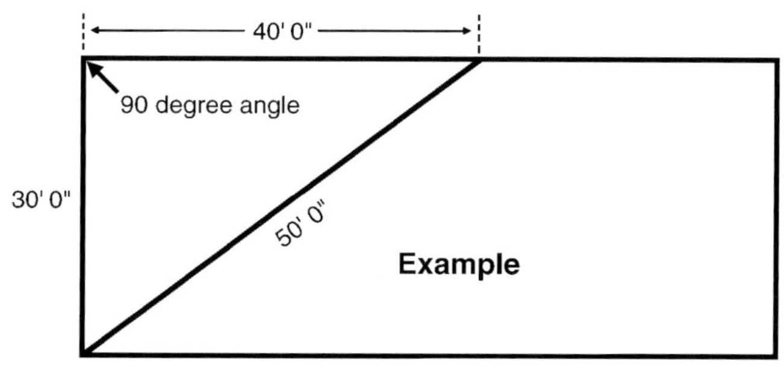

Things get more difficult when your tunnel is not a multiple of the 3-4-5 triangle. For example, using the Theorem, we find the diagonal on a 20' x 96', (which is a common size), is 98.06 feet, or 98'- 5/8". Measuring that distance out across the diagonal and marking it will give you a 90 degree angle for this size house. The opposite diagonal will measure the same when your layout is square. If all this is too much, and you are purchasing the materials from a greenhouse supplier, ask them for the diagonal measure.

Once the corners are marked, install the batten boards about two or three feet out at each corner. Pull strings across the corner stakes at exactly on-center on all four sides and attach them to the batten boards. A screw drilled into the spot on the batten board where the string is to run will help you draw it tight and attach it securely. The tops of batten boards should be approximately 6 inches above grade. Draw the strings tight , and use the string bubble level to insure that the whole house is level. Pound the batten boards in farther or pull them out a bit until your bubble level gives you the reading you want.

After you have established the center line for the perimeter of the house, at the 6" level above grade, you will need to mark the position of each ground stake using a marking flag or large nail. If the columns are 4' apart, you will place the flags every 4' down each side. If your house has 6' column spacing, do the same at 6'. Once the locations for each stake is marked, move the strings outward on the batten boards half the diameter of the columns or hoops you will be installing, as illustrated below.

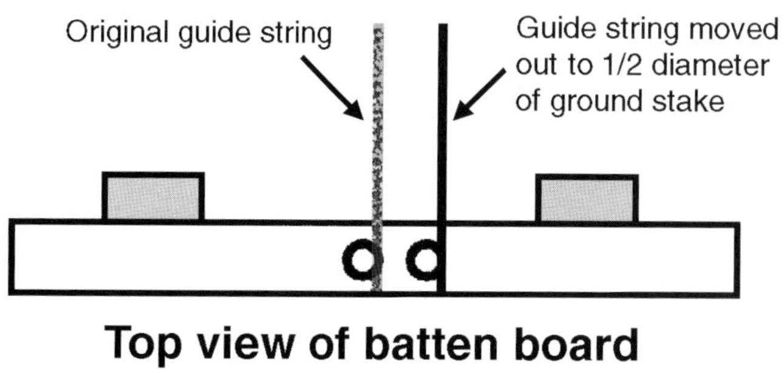

A small nail driven at that distance on each board will make this easier. If all goes well the strings will now mark the outside edge of the house and a ground stake will be driven at each flag with its outside edge just touching the string. If the strings are leveled as you want, you will drive the stakes to where their top just touches the string on their outside edge. See illustration on the next page.

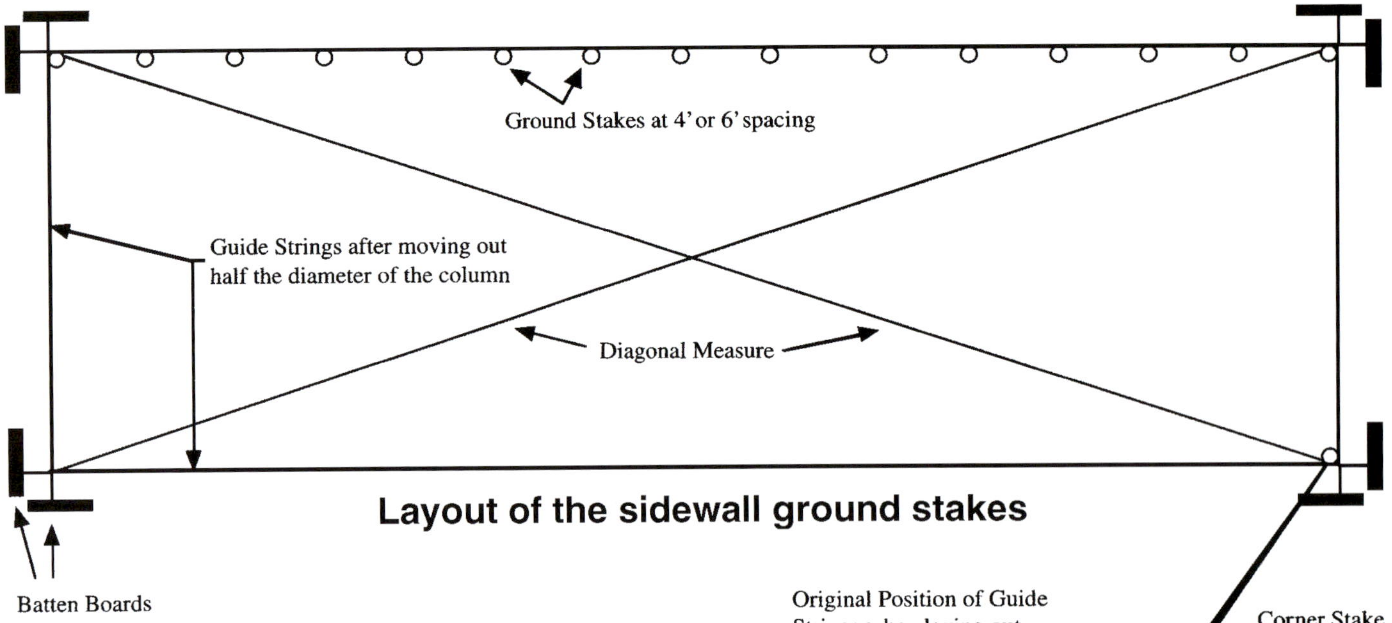

Layout of the sidewall ground stakes

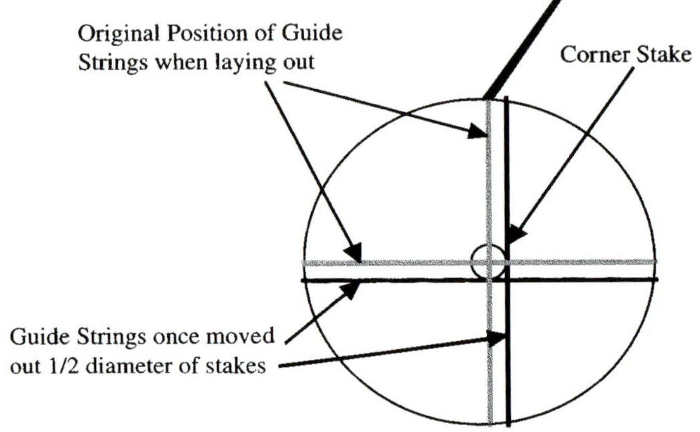

Building the tunnel

If you purchased a high tunnel from a manufacturer, it should come with assembly directions. Unfortunately, we have heard from many growers who said they didn't understand the directions provided. In this section, I will explain the basic procedure in the hope that it will illuminate the manufacturer's directions as well as help those who are building from scratch.

Lay out the ground stakes and begin driving them at each flag with a sledge hammer, taking care not to damage the top of the stake. Inserting a trailer ball hitch into the top of the stake to absorb the hammer blow will protect it. Make sure the stakes are straight up and down. An alternative method is to dig post holes at each flag, and set the ground stakes into concrete. In windy areas, at least the corner posts should be set in concrete.

Once all the stakes have been driven, you may begin erecting the hoops. Most manufacturers make the hoops in several pieces that have to be assembled before you put them on the posts. A cordless drill, magnetic socket and self-tapping tech screws with bolt heads will ease assembly. Be sure to assemble them on a flat surface.

As each hoop is completed, two people will be needed to carry it to the far end of the house and place it over the swedged ground stakes on either side. Be aware that the ground stakes are placed slightly closer than the natural spread of the hoops, so they will need to be compressed in order to slip over the stakes, but this adds structural rigidity. Attach the hoops to the ground stakes with self-tapping tech screws. Work your way forward.

A typical hoophouse kit as it arrives from the manufacturer.

Once all the hoops are in place, begin attaching the top purlin at the peak of the house. To insure the 4' or 6' distance is maintained between the hoops, it is useful to make a wooden template with which you can snug each hoop up before attaching it with tech screws. See illustration below.

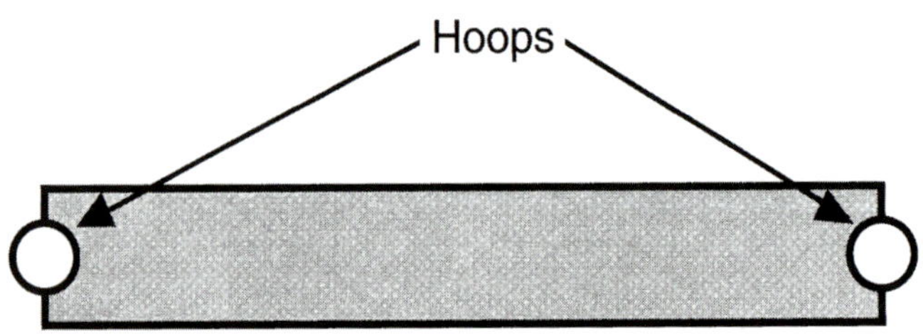

4' or 6' template board that fits between hoops

The purlin is often made of the same pipe material as the hoops. Be sure to screw each length together at the swedge to prevent it from coming apart.

After the ground posts are placed, hoops are assembled and then inserted onto the ground posts, above. Once all the hoops are placed, one or more purlins are screwed to the hoops to connect them. A board cut to the correct size serves as a template that will ensure correct spacing of the bows when screwing on the purlins, at right.

For this hoophouse, locally purchased lumber was used to make hip boards and base boards. An alternative is to purchase a kit with aluminum channels for hip boards.

Frame in the end walls to accommodate whatever kind of doors you plan to use. This hoophouse has double doors that open outward, allowing tractor access to the beds inside.

Normally there are angle braces at the ends of the house between the first and second hoops. These should be attached with tech screws.

Hip and base boards

You are now ready to begin attaching the hip boards. If your house has straight sides, as opposed to a ground-to-ground arch, attach the hip board at the top of the side. Treated 4" x 1" lumber can be used for the hip boards because it will not come in contact with the soil. The template you used for snugging the purlin will come in handy to assure the same distance is maintained at the side wall. Attach the hip boards with larger tech screws. Tie the hip boards together with flat metal truss plates.

Most hoophouses can be purchased with aluminum channels for attaching the poly on the sides. A less expensive option or for DIY hoophouses is to attach a 2" x 1" furring strip at the top of the hip boards he entire length of the house. This will serve later to help draw tight and secure the plastic covering.

Attach base boards down the entire length of the house. We use 6" or 8" lumber that also serves to retain soil in the house, and keep out grass and weeds. It is a good idea to install woven weed barrier over this board and out a couple of feet away at this time.

If you are concerned about wind strength, you can go back and dig around the corner posts and possibly a middle post on each side and pour concrete to further anchor the tunnel at this time.

You can now install the end walls and system for attaching the covering over the hoops at each end of the house. End walls are as different in conception as there are builders. Some prefer a closed, wooden or polycarbonate end wall. Others prefer a removable one-piece end wall. Still others prefer large doors that can be swung open to admit a tractor or allow heat to escape. In snowy places, inward opening or sliding doors may be necessary to gain access in winter. Some growers opt for simplicity and low cost with boards or pipes holding the plastic curtain and swinging down from the top center of the hoops, to attach at ground level in the center of the house. We prefer a framed end wall with large enough opening to accommodate a tractor, but we entirely remove the plastic covering from the end walls during the summer. To attach the covering over the hoops, we use an aluminum channel with a high tensile zig-zagged wire, called wiggle wire, which locks the poly into place. There are several other systems that serve to hold the poly in place, and the one you use will probably depend on which greenhouse company you buy from.

Once your hip boards, end walls and wiggle wire channels are in place, you are ready to cover the house with greenhouse poly.

Covering the tunnel

First of all, don't use cheap plastic from the hardware store, which is not UV treated and will break down quickly, maybe before your first growing season is up. Use

Prepare the furring strips that will be screwed onto the hip boards to hold the plastic in place. Have the screws started in the strips and the strips laid out along both sides of the house.

greenhouse poly. Quality is counted in seasons of use, with most poly expected to last four years. Make sure your poly is wide enough to accommodate your tunnel; the supplier can help you calculate the amount you need to cover the hoophouse and leave some excess with which to attach the rolled up sides.

Install the poly on a calm day. You will need at least three people for a full-sized house, and one more person would be helpful. Prepare by having ladders in place, presink screws into the furring strips with which you will attach the sidewalls, and pre-position wiggle wire or other poly locking components.

Some plastic comes folded in the middle and can be unrolled down the peak of the house. In most cases, though, you will be unfolding the plastic on the ground and then pulling it up and over the arch.

Unroll the plastic parallel to the house, making sure there is a good overlap on each end. Find an edge on the top of the laid out plastic (one on the bottom will mean the plastic will have to flip over as it comes up). Attach baling twine or rope to the edge by bunching up a handful of plastic or surrounding a tennis ball, and tying a knot around it. Throw the rope over the top of the arch.

At least three people should pull the plastic gently over the arch with the ropes, while one person makes sure it unfolds freely and does not bind anywhere. Once the house is covered, make sure it is square by siting the fold lines down the length of the house. The center fold line should be at the peak and the plastic should hang equally on both sides with a good overlap on each end.

Attach the poly with the wiggle wire (shown at left) or

Unroll the poly along the length of the house and spread it out partially. Gather it up and tie a long piece of rope around it, at the ends and in the middle. Throw the ropes over the top of the house, and have people on that side pull the plastic gently over the hoops.

Attach the poly to the ends first, trying to get it as taut as possible. If there is no breeze, take your time with this step and get the poly centered and stretched. But if there is any breeze at all, just get it attached quickly and plan to stretch and reattach it later.

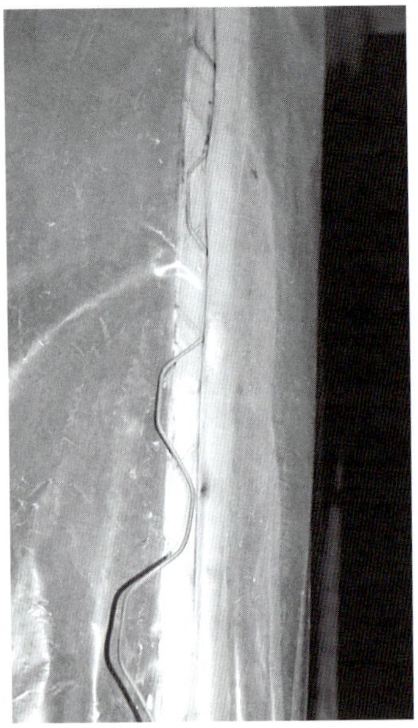

other attachment system, beginning at the corner from which most wind is coming. Completely attach one end, then the other. The poly is fairly secure now and you can begin attaching it at the hip boards on the sidewalls.

You should tighten the poly as you go by drawing it toward the far end while installing the furring strips (shown below) or other attachment system. One person might draw the poly while a second assists with snugging the strips up and the third operates the cordless drill, attaching it. Alternate in this from side to side as you move from one end to the other. When you reach the far end, you will need to remove the wiggle wire, draw out the slack you have gathered, and replace the wiggle wire.

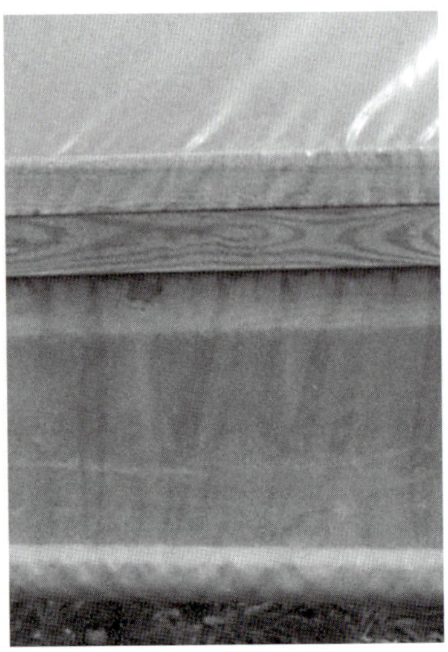

Be aware that on a warm day, the plastic will appear loose and on a cold day it will appear drum tight. You may back out the screws on the furring strips and draw the plastic tighter on a warm day if you wish.

You now have plastic covering the arch, but

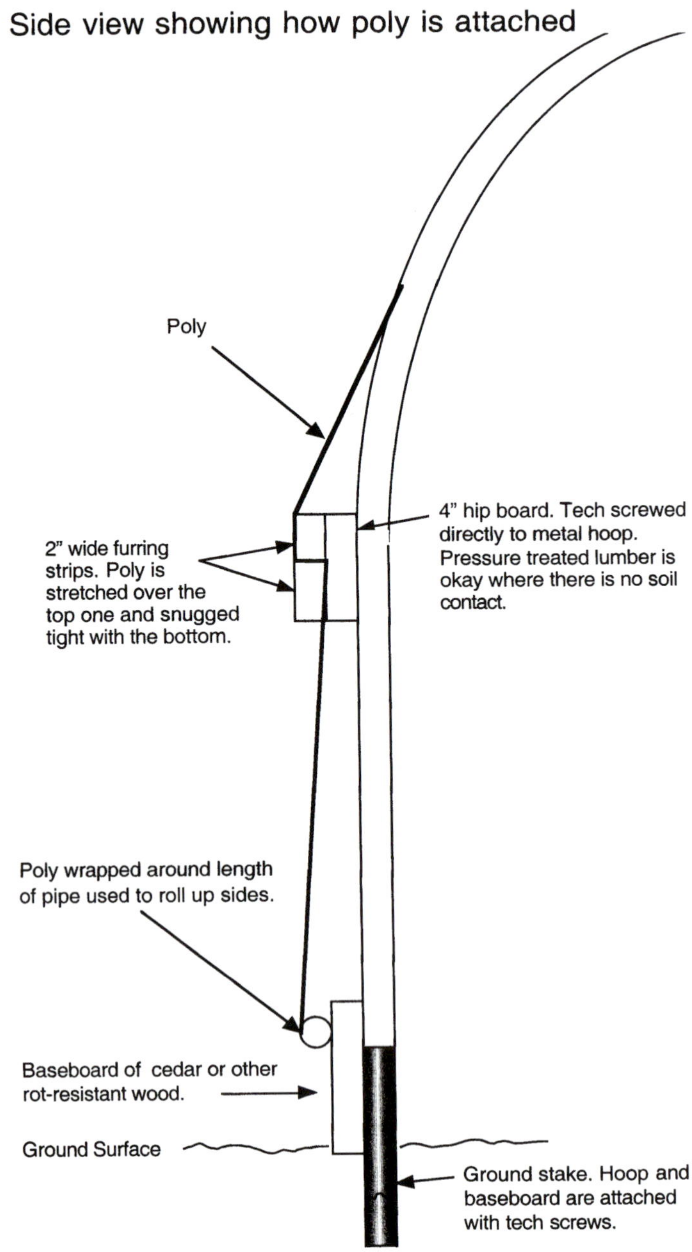

loose and flapping below the hip boards. To create roll-up sides, run pipes the length of the house and wrap the plastic around it. You can attach the poly to the pipe with tech screws and strips of used drip tape, which prevents the screws from tearing the poly, as shown above.

There are various ways to make a crank at the end of the pipe. See page 48 for photos. You will find it is surprisingly easy to roll up the entire side of the house with a few simple cranks. Roll the sidewall up and hold it in place. Attach webbing over the sidewall from the hip board to the baseboard using wood screws. We find that used drip tape works very well for this, as shown at right.

All that remains now is to cover the end walls. Regardless of the structure, if you cover them with the same poly, you can use the same aluminum channel that holds the roof in place. The channel will accommodate a second covering and second wiggle wire. This allows you to take off the end walls without disturbing the roof.

Your hoophouse is now ready for production!

How to roll up the sides

These photos illustrate three different approaches to roll-up sides. Top left, an L-shaped handle made of PVC pipe and connectors is screwed onto the metal pipe on which the poly film is rolled. A rope or chain attached to the endwall frame can be slipped over the handle to hold it in the up position. Top right, a longer piece of PVC pipe slides through a connector and props the sidwall open. It's more cumbersome because of the length of the pipe, but it's cheap and easy.

Bottom photos: University of Missouri researchers came up with this design. Lewis Jett demonstrates rolling up the sidewall with the short handle; the longer piece of pipe then slides in and holds the sidewall in place.

Greenhouse manufacturers also make crank systems for rolling up or down the sides.

Doors and endwalls

Doors can be as simple or as complex, as cheap or expensive, as you wish. At left, a poly-covered frame on a sliding rail creates a wide door.

Below, a purchased storm door makes for a tigher closure and neater appearance, but is too narrow to move carts or tractors into the hoophouuse. However, large sections of the endwall can be removed for equipment and summer ventilation.

Bottom, two frames covered with poly open outward, creating an entry large enough for a tractor. Inward opening doors may be preferable in snowy areas but they will require room to swing inside, eating up valuable growing space.

Drop-down curtains gain popularity

Most hoophouses are built with side curtains that roll up for ventilation. That's an easy and inexpensive way to regulate temperature without the use of fans and vents. In most designs, the roll-up portion of the side is simply an extension of the poly that covers the roof of the tunnel. The bottom of the poly is attached to a pole with a crank at one end, and one person can easily roll up and roll down the sidewall covering.

But drop-down curtains are gaining in popularity. Norman Kilmer, the owner of Morgan County Seeds, says that every high tunnel except one that he has sold in the past couple of years has been ordered with drop-down curtains. Other manufacturers, including Farm Tek and Atlas Greenhouse Systems, are also offering drop-down curtains on their newest high tunnel models.

When a roll-up sidewall curtain is opened, outside air pours into the tunnel at ground level. In winter and early spring, the flow of cold air onto the soil can harm young plants. Wind can also get into the tunnel and damage growing tips of plants.

In contrast, a drop-down curtain allows air to flow in at the top of the sidewall, usually 4-6 feet above the ground. The soil stays warm, while excess heat rises up and flows out above the curtain. The plants are protected from the wind, too.

"If you're growing a crop you want to keep cool, you might want a roll-up curtain," Kilmer said. "But if you're growing in a high tunnel to provide heat to the crop, a drop-down curtain is a better choice. And in summertime, you can drop it all the way down."

A drop-down curtain on an Atlas Greenhouse Systems tunnel.

Drop-down curtains have long been used in poultry and hog barns and the system was easily adapted to greenhouse poly for high tunnels. The system uses steel cable, pulleys, and ropes to drop and raise the curtain.

Most high tunnels can be retrofitted with the drop-down curtains, Kilmer said. He sells all the components for a drop-down side curtain on a 96-foot-long tunnel with 5-foot sidewalls for $509. "It's not that high priced, and it's quite simple to put them on," Kilmer said.

Automatic openers

For growers who can't be at the farm to manage ventilation needs, there are automatic openers. The photos at left and below show the the Poly-Vent, an inflatable sidewall that deflates when a specified temperature is reached inside a hoophouse. As the temperature drops, fans turn on that inflate the sidewall. It is available as a one-stage system that fully opens or closes the sidewall, or as a two-stage system that provides partial ventilation. Cost for a 96-foot-long house is about $2,000, depending on options. poly-tex.com.

Prevent problems with design

Commercial greenhouses are built to specific structural standards so that they can withstand snow, wind, and things that might be hung from the structure such as hanging baskets, heaters and fans. That strength is the reason for the expense of a greenhouse.

A hoophouse or high tunnel, in contrast, is generally what's called in the greenhouse industry a "non-code house," meaning it doesn't meet building codes for wind, snow and other loads. That's why it's so inexpensive.

Even so, hoophouses can be made strong enough to withstand most severe weather. Unfortunately, without codes, growers may have to figure out through trial and error how strong is strong enough. Many hoophouses have collapsed under ice, snow, or wind and the growers have recognized in hindsight what they should have done differently. In the absence of standards for hoophouse construction, we went in search of anecdotal information about how to keep the hoophouse standing strong.

Snow loads

Wet snow and freezing rain can quickly add thousands of pounds of weight that can crush the purlins and bows of a hoophouse. Growers in extremely snowy climates often take the plastic off in winter to avoid the problem. Some use emergency heaters to heat the air inside the house so that the snow melts when it hits the warm poly.

But when those aren't an option, growers must choose structures with sufficient strength. Thicker steel is the best option - look for pipes with an outside diameter of at least 1.66 inches. Steel quality can be an issue, too, which is why you should check to see if the manufacturer is a member of the National Greenhouse Manufacturers Association, which sets quality standards.

An ice storm brought down these cold frames when the single purlin bent under the weight of the ice, in turn bending the hoops through much of the house.

Adding more than one purlin can strengthen the house, as can additional cross bracing.

The shape of the house can have an effect, with the peaked roof of a gothic arch or gable structure more able to shed snow than a quonset shaped house. Houses with higher sidewalls are less stable and require stronger posts and hoops that those without sidewalls or with short sidewalls.

Wind loads

One greenhouse sales rep described the hoophouse as "a big kite you need to anchor to the earth." The features mentioned above for snow load can also make the structure more resistant to wind, but additional precautions may be needed to counter wind pressure on the sides of the structure. Wind damage can occur in several ways:

•Straight line winds or gusts can hit the end walls and push them down if they aren't strong enough. When the endwall falls inward, the purlin breaks or bends, often harming the bows as well.

•Wind can hit the side of the hoophouse and be forced upward, where it contacts straight winds above, which creates a churning effect that can tear up hoophouses behind the one closest to the wind direction.

•Wind that gets inside a hoophouse, through an open door or though roll-up sidewalls that have blown outward, can actually lift the structure out of the ground.

Steve Upson at the Noble Foundation in Ardmore, Oklahoma, has used several strategies for making his hoophouses more able to withstand wind. His end walls are built with four posts, each made of a double 2x6 bolted together and sunk 36 inches in concrete. He also uses 2-inch poly straps to restrain the side walls from blowing outward, rather than the drip tape that many growers use for the purpose. He also uses a sidewall curtain of shade cloth to serve as a windbreak when the sides are rolled up.

Many growers use webbing straps across the top of the hoophouse to secure the plastic.

In a storm, the best strategy is to close the hoophouse entirely so that wind can't get inside and pull the house upward.

Orientation

Although orientation of the house can affect how the wind hits it, greenhouse manufacturers recommend that you orient the house to get the best light, and let structural modifications take care of wind load. The rule of thumb for orienting the house to capture the most light in winter is that the greenhouse should be oriented with the ridge running east to west for locations north of 40° latitude and north to south for those located south of 40° latitude. The

With the endwalls open in summer, wind got inside and lifted this tunnel's posts out of the ground, bounced them around and bent the purlin and about a third of the hoops.

40th parallel of latitude crosses the United States in a line that stretches from Philadelphia, PA, through Springfield, IL, the Kansas-Nebraska border, Denver, CO, Provo, UT, and Fort Bragg, CA, as shown on the map below.

The reasoning for this general rule about greenhouse orientation is that in the north, the sun is closer to the horizon in winter; it's best to have the sunlight enter along a side of the greenhouse rather than an endwall, where it would be blocked by the frame. At latitudes below 40°, the angle of the sun in winter is higher and won't cause shadows from the endwalls.

At any latitude, gutter-connected or closely spaced multiple greenhouses will get more light if they are aligned north-south because they avoid the shadow cast by the structures south of it. The difference in light intensity is not great during the summer when the sun is high in the sky, but there can be a big difference in winter. Research done in England, at 50°N latitude, showed that the percent light transmission in midwinter was 71% in an east-west greenhouse, and only 48% in a north-south greenhouse.

Multiple houses

If you are going to build more than one hoophouse, be sure to build them far enough apart to allow good air circulation and prevent shadows in winter if you will be using it then. Look at the angle of sunlight on the shortest day of the year, December 21, to determine the length of the shadow cast by the southernmost hoophouse, other structures or trees. The farther north you are, the lower the angle of the sun, and the longer the shadows, which means wider spacing between houses.

Another consideration in spacing multiple houses is the amount of snow that you get. Leave room for snow to spread out beside the hoophouse when you brush it off. If it has to drop off into a narrow alley between houses, it will pile up and put pressure on the sidewalls that could actually bend the posts. In addition, it could take a really long time for that pile of snow to melt.

If building multiple high tunnels oriented east-west, the spacing between the tunnels needs to be twice the height of the tunnels. If they are oriented north-south, the spacing can be 4 feet between tunnels.

Drainage and weed control

In you have heavy soil that tends to hold a lot of water, or if you have water cutting across fields, you may need to install a drainage system around the house. A trench with a drainage pipe covered with gravel will prevent water from running into the hoophouse, and it may also prevent weeds from encroaching into the hoophouse from the sides.

Even without drainage problems, it's a good idea to use heavy landscape fabric around the perimeter of the hoophouse to stop weeds. Use the permeable woven kind, so water doesn't puddle on it.

DIY hoophouses save on costs

Steve Upson of the Noble Foundation in Ardmore, Oklahoma, says that "Spanish" style tunnels are popular in the South and can be constructed for less than $1 a square foot. At left is a tunnel built by Tod and Jamie Hanley of Norman, OK, using a pipe bender to make the hoops themselves. Instructions are on the Kerr Center web site at www. kerrcenter.com/publications/hoophouse.

Below is a tunnel designed by Upson with hoops of industrial poly pipe commonly used in ground-source heat installations. The poly pipe does not require a bender. The Noble structure is equipped with a split vent that allows venting without exposing plants to cold air.

Both the Hanley and Noble Foundation models use rope to attach the poly to the hoops.

At left are farm-built tunnels on a cut flower farm in southern California. The minimal covering protects flowers from wind, sun, and rain damage.

PVC hoops

Anne Cox built several hoophouses out of PVC pipe to give tomatoes and other warm-weather crops a boost in her cool Maine climate. She has three farm-built tunnels. The upper one is 14' x 34' and the lower two are 12' x 28'.

The PVC pipes are attached to the wooden raised beds at the sides to make hoops 2 feet apart. The ridge pole and purlins are bolted to every hoop.

"A few other details: I put the posts down the center supporting the ridge pole as temporary winter snow load supports (am I ever glad they were up this winter!), but decided to leave them in the tomato hoophouse to deal with the load of the vine trellising I attached to the ridge. And the pavers on the bottom were just something that came from a landscaping project (my other hat), and I thought they might help as a heat sink in cooler weather. I think they help (and they look sweet)."

Johnny's Quick Hoops

Johnny's Selected Seeds sells tools and materials for making several versions of Quick Hoops tunnels. Shown below is the Gothic Tunnel, which can be built as a stationary tunnel or a modular, movable tunnel. Pipe benders are available for Quonset-style high tunnels and low caterpillar tunnels. Johnny's sells the specialized parts necessary for the tunnels (many of the materials needed are locally available) and has detailed instruction manuals and cost calculators available free on their website: www.johnnyseeds.com. Search for Quick Hoops to see the complete selection.

Hoophouse ventilation ideas

One of the biggest management issues for hoophouse growers is temperature control. A closed-up hoophouse becomes an oven as soon as the sun hits it, on even the coldest days, and must be ventilated. Good air circulation is critical for plants year-round. Ventilation is critical to using hoophouses year-round.

In Eagleville, Tennessee, Jim and Julie Vaughn designed windows in the endwalls to increase ventilation. "We have three 35x144 passive solar hoops," Julie said. "My husband came up with the idea for these windows to ventilate the "tops" of the hoophouses since these houses are so tall. We made a window frame out of extra pieces of end wall metal, covered with corrugated sidewall covering that overlaps the endwall window frame to form a tight seal when closed. He added home-made PVC weights to help the windows descend when opening. In order to open we have a winch system set up inside the houses, one for each window. The winches have cranks and catches to keep the windows at the level you want them. These windows are on both ends of our hoophouses. In our winter growing months this is usually the only ventilation we use. The windows really help to let the moisture out of the houses without allowing cold air to enter the houses at plant level. The great thing about our windows is that you can open them as much or as little as you want/need, which is great for the extreme weather days (hot, cold, rain) that we can encounter here in TN. It is also a great "safety" for us to allow some ventilation ahead of storms without getting caught with our sidewalls down."

Jim and Julie Vaughn ventilate their tall hoophouses with windows in each endwall, which can be opened and closed with the crank at right in top photo. Homemade PVC weights on the outside help the windows descend.
Below, sliding interior doors keep the Van Sandts' hoophouse accessible in deep snow, and sliding fences keep out animals.

Beth Van Sandt of Scenic Place Peonies in Homer, Alaska, chose a ventilation system that will work in her snowy winters.

"We opted to go with inside sliding doors on both end walls for our system of ventilation. Due to our ginormous amount of snow in the winter we opted for inside sliders so as to be able to get into the HT. Each door is 4x8 which leaves us a nice large opening at each end for air flow. We built movable fencing for each end to keep out our free-range chickens and snowshoe hares."

The solution pictured on this page is from Kathy Thorne of Blue House Market Garden in Loranger, Louisiana., who keeps the sidewalls rolled up all summer in her hot climate. She wrote: "The system uses pipehangers ($0.90 each) and 1-inch aluminum channel ($12 for a 3-foot length). The ends are tucked under and clipped on the end ribs. I anticipate that it will be easy to stow when I need to roll the sidewalls down - just pull the pin and drop the hooks."

A quick and easy way to cover hoophouse crops

By Adam Montri

As temperatures drop and day length shortens, it's time to get our interior covering in place inside the hoophouse for the coldest months ahead. With an interior covering, or what we also call the "inside tent," or the "hoops within the hoops," it is possible for growers in the North to harvest year-round. Farmers farther south may not need the interior covering, but for us in Michigan and similarly cold climates, it would be challenging to harvest past December without it.

How it works

Just as you would add another blanket on your bed to stay warmer in winter, you can add a blanket on your crops to keep them warmer. Another way to think about it is to consider the way the hoophouse works. We know that it is warmer on a cloudy night than it is on a clear one. Why? Because the clouds act as an insulation layer to trap the Earth's radiant energy from moving off into space. And what are clouds made of? Water vapor. When we hit the dew point, water condenses on the coldest surface. In the hoophouse this is the polyethylene layer. When this happens, the hoophouse more or less acts like a cloud trapping the radiant energy (heat) that is coming off the soil surface that has warmed up during the day. In the spring and fall, the hoophouse covering itself is enough to trap that radiant energy, or at least slow down the escape. But in the winter, there is an opportunity to reduce the space between the crops and the polyethylene layer. Even the tallest crops (kale, chard, collards) are only three feet tall. By adding an interior covering sometime in late October or early November we can keep the temperatures warmer closer to where the crops are planted.

Choosing a cover

The two most common materials for the interior covering are row cover and polyethylene. Both have benefits and challenges. We use row cover (Agribon-19) that weighs .55 oz. per cubic yard. There are a number of companies that make this weight and we choose this one based on balancing enough light coming through the covering and enough

An inner cover allows year-round production at the author's farm in Michigan, and the high-tensile wire system he devised makes it easy to cover an entire hoophouse full of winter crops. Photos by Adam Montri.

weight to trap the heat. That being said, let's talk about the poly option before coming back to why we use row cover and how we install it.

What is nice about the poly covering is that it can be old poly from when you change the roof poly. On sunny days it is pulled back so even if the light infiltration is low (why it was taken off the roof in the first place), it is not going to inhibit light as long as it is managed well. It is also more durable than the row cover — you can be a bit rough with it and not have to worry about tearing it, which can more easily happen with row cover. It also tends to be a bit warmer than the row cover. The two downsides to the poly are that it is not breathable and it is heavy. This means that if you are not always on the farm to pull it back to vent on a sunny winter day then it gets very hot, very fast, underneath it. Mismanagement of the poly cover makes the temperatures way too hot to keep things like spinach, lettuce, salad greens and other winter crops happy for very long. Because it is not breathable, it is more humid than under row cover. Those of us who live in humid places know that high humidity and warm temperatures add up to high chances for fungal diseases. The weight of the poly also means that it needs more support to keep it above the ground. This translates into higher cost to install it.

We talked about the disadvantages of the row cover – mainly that it is not as durable and potentially not as warm as the poly. The benefits of the row cover is that it is breathable and lightweight. That means that if it gets sunny and we don't open the cover, the crops do not overheat. It also means that it is not as humid, which reduces the opportunities for diseases. On our farm, we both have off-farm jobs related to agriculture. In the winter, this means we leave before the sun comes up and get home after it goes down most days. There is no way we are going to open the cover when it is dark and -15°F at 6 a.m. because it is going to be sunny later in the day and then try to close the cover when it's dark and it has frozen to itself. We learned that one the hard way. Also, since it is lightweight, it takes less support to keep it above the crops which means lower investment costs.

Building the support

There are a number of options for supporting the row cover inside the hoophouse. There are wickets, hoops of either PVC or EMT conduit, #9 wire and a number of other ways. There is also the option to either cover beds individually or to cover large areas or the entire growing area depending on how the beds are arranged. Since our overall goal is to trap the radiant energy close to the crops, we decided to cover the entire growing area. Using the analogy of a mitten versus a glove helps us explain why covering the entire growing are is warmer than covering individual beds. Is a mitten or a glove warmer? Most of us in the North know that mittens are warmer than gloves because there is less surface area to lose heat from in relation to the area covered.

We wanted to come up with a different interior covering design that allowed us to cover the entire area but also had minimal support investment. Working with John Biernbaum from Michigan State University, we designed a system that used high-tensile wire to support the row cover.

The wire system with row cover drawn across the entire width and length of the hoophouse. Note the post at left, where the wire is attached.

We have metal endwall framing covered with polycarbonate on our 34' x 96' hoophouse and our beds run the short way across the house. They are 3' wide with 1' walkways between them. We also have two 4' x 8' sliding doors (mounted on the inside so we don't have to shovel snow to open them) on one end so that the door, when fully opened, is 8' x 8' We have a number of vertical uprights as part of the endwall framing, but the important ones for this are the two that are on each endwall just at the outside edges of the beds. They are directly opposite each other end-to-end so that the high tensile wire can be run in a straight line the length of the house and attach to each of these uprights on the outside of the growing area. These are two of the four wires that we run. For the middle two wires we could not attach them to the endwall because the verticals there are part of the door frame. If we attached them there, they would block the doors from sliding open and we would have no way to get inside. For these two center runs of wire we cut two 7' pieces of leftover endwall steel (2" square tubing) and buried them 3' in the ground with 4' above ground. On the near end we installed a wire strainer for high tensile wire (available at the feed store where we bought the wire) on each of the verticals. They are designed to bolt to a vertical and can be attached by drilling through the steel and using a 3" or 4" nut and bolt. If we had wooden endwalls we would not attach them directly to the endwalls because of the potential strain. Instead, we would bury posts and attach to them as you would for animal fencing. At the far end, we bolted tension measuring springs that tell us the pounds per foot of tension on the wire using the same idea of drilling through the endwall vertical. We don't necessarily use the

Above, the hoophouse with the cover partially open. Note the clothespins on outer wires and how row cover drapes to the ground. Left, a wire strainer attached to the endwall. Below, a tension measuring spring attached to the opposite endwall (viewed from above.) Photos by Adam Montri.

pounds of tension to set the wires as much as the springs make a good attachment point. We then attach the wire and tighten down the strainers so that there is no sag in the wires. That is the only support we use for the entire area. The wires are approximately 40 inches above the ground.

Covering

For the covering we use three pieces of 30' x 40' row cover. We bought two rolls of 30' x 100' and cut them to this size. We could have bought one large piece but we chose to use three smaller pieces that overlap because that way if we want to get to the middle bed we don't have to pull from the far end all the way down to open it. It also makes opening the cover on a sunny day fairly easy by pushing the two outside pieces to the ends and bunching the middle piece over an aisle. The 40' runs the short distance across the house. This way, the same piece runs from the ground on one side up and over the 30' long beds and down to the ground on the other side. The 30' length of the pieces runs the length of the house and the three of them, when just slightly overlapped, seal each other and total the length of the production area. These are all held in place with clothespins every 8-10' on the two outside wires. We pull the row cover tight before putting the clothespins on so there is no sagging in the row cover. We especially want to make sure that the covering does not touch the plants because when the water that condenses on the covering freezes, it will damage the plants if it is in contact with them. At this point, the tops and the sides of the area are covered but the ends are still open. To close these we cut two 5' x 30' pieces of row cover (one for each end) and attach it to the wires using clothespins. We clothespin one corner to one of the outside wires and then pull it tightly as we go across the house, securing it to each of the wires with a clothespin. We don't move the ends once they are in place.

Cost vs. revenue

Materials for a 30' x 84' x 40" cover:

400' 12.5-gauge high tensile wire = $10

4 wire strainers @ $3 each = $12

4 tension measuring springs @ $7.49 each = $29.96

2 row cover pieces, Ag-19 or similar (cut to three 30' x 40' and two 5' x 30') = $140

1 pack clothespins = $5

TOTAL COST: $196.96

Note: We attached wires to the metal endwall verticals and two pieces of leftover pipe. If posts need to be purchased, there would be an additional cost.

This covers 20 beds in our hoophouse. With one 3' x 30' bed of salad mix cut three times we can generate around $450 (gross) when sold at $10/lb. Based on this, payback is not a concern with this covering.

Management and maintenance

It is important to make sure that there are no gaps in the covering. This means that it reaches the ground all around, and that where the top pieces overlap they are pulled tight and clothespinned together. To make sure there are no gaps at the ends, we drape the top pieces over the end pieces by a foot or so, depending on how much overlap there is. It works very well with two people to just slide the row cover along the wire to open it with one person on each side. It is faster with two people but with one it really isn't a challenge, it just takes 10 minutes instead of five to open it up. Ideally, we would like to pull the cover back whenever it is sunny to be able to get additional heat into the soil, but in all honesty we do this relatively little in the winter since we are off-farm as mentioned above. When we are home we do pull it. Opening the cover will decrease the humidity in the microclimate surrounding the crops and therefore decrease the chance for disease such as Cercospora leaf spot on spinach and beets and white molds on tightly planted lettuce for baby leaf salad mix.

While it could be in one single piece we also like the three pieces for management. Since we can only harvest on days that the temperatures reach above 32°F (freezing), we like the option to move only a portion of the cover when we harvest. With the three separate, overlapping pieces we only have to move the cover a maximum of 3 beds in either direction to access the production area. The height at approximately 40 inches is not only meant to accommodate the taller plants but also allows us to crawl underneath, if need be, to harvest without opening the cover. We do this a few times a year when it is just above freezing, but cloudy, so we don't want to open the cover and let out all that built-up heat.

We also like the high-tensile wire because it can be rolled up and coiled (with some patience and maybe some choice words!) and stored in place for the summer. The cover can be rolled and hung from the rafters or folded and put in a sealed container to prevent mice from nesting in it. Our current row cover has been through two winter seasons and a portion, if not all of it, will be replaced this year. Lastly, as we move into April and our tomatoes and other warm season crops start to be transplanted into the hoophouse, the cover protects them as well. This is where the only challenge with this covering has come up. Our roll-up sides are also thermostatically controlled. Often in Michigan, like other northern states, our early spring days are cold in the morning but can be sunny and warmer as we go through the day. There have been a few times we have come home, the sides have rolled up and the row cover is blown around a bit. It has been nothing major and not damaged any crops but does take a bit of time to unwind and get it back in place.

Adam Montri owns Ten Hens Farm in Bath, MI, and works with farmers through his outreach position with the Michigan State University Student Organic Farm. For more information, visit www.hoophouse.msu.edu.

Common hoophouse problems — solved!

By Ben Hartman

Like many fresh vegetable farmers, we continue to explore the possibilities with hoophouses and greenhouses. We have two greenhouses that we constructed within the last two years. One measures 20' x 70', the other 30' x 90'. We are in northern Indiana and have a CSA that runs June to mid-December, and we sell at our farmers market all year round, so we use our greenhouses a lot. Here are three problems we encountered and solutions that have worked for us.

Problem: Ventilating when we are not at the farm.
Solution: Automated shutters and end wall fans.

Our usual routine with our first hoophouse from late fall to early spring was to open up in the mornings and close up in the evenings. But what if we weren't at home? More than once we found ourselves rushing home from town to open up the hoophouse once the sun unexpectedly came out.

At an auction last year we purchased an assortment of exhaust shutters and fans. Some of the shutters came with automated motors attached, for others we purchased add-on kits through Nolt's Produce Supply in Leola, PA. On the south ends of both greenhouses we installed two exhaust fans, on the north ends we installed three shutters — two lower and one upper. We also installed recirculation fans to keep air moving inside the greenhouses.

In the middle of each greenhouse we set up a control panel, to which we mounted corrosive environment thermostats, one for each device. (See photo on page 12.) The thermostats cost around $30 each. We admit that the electrical work was beyond our expertise: we traded a CSA membership for help from an electrician friend.

While wiring these devices does require special knowledge, the concept is simple: For each device, we simply set the thermostat to the temperature at which we want the device to kick on. We usually do NOT want all the devices to come on at once, so we adjust the thermostats so that they come on in stages. Our aim is to keep air at plant level under 75°F while minimizing plant damage caused by rushes of cold air. The automated system allows for much more accurate control of temperature than our old manual systems. And we don't have to sweat when we're away from the farm when the sun comes out!

At right is an example of our settings last winter. Note that as the indoor temperature increases, more devices run simultaneously.

Temp at control panel	Device powered on	Result
65°F	upper shutter (north wall)	Lets hot air escape without cold air rushing on plants.
70°F	recirculation fans	Mixes cold/hot air in the upper chamber, away from plants.
75°F	left endwall fan (south wall)	Pulls cold air from upper shutter. Cold/hot air mix in the upper chamber before being exhausted out. This was often enough ventilation in the winter.
80°F	right endwall fan (south wall)	Doubles exhaust strength. Mainly for summer use.
85°F	lower shutters (north wall; wired to one thermostat)	Allows maximum ventilation. This can pull cool air across plants on cold days.

We have seen many different ventilation configurations and control panel settings, and we ourselves are still experimenting. What's important is to create the most ideal temperature environment for the specific crops you are growing. Your greenhouse supplier should have data available to help size an automated exhaust system for your location/size of greenhouse.

To save electricity, we still use roll-up curtains and manually open and close doors when we can.

Problem: Roll-up curtains allow cold air to rush across tender plants.

Solution: Short walls behind roll-up curtain.

The short walls pictured at right are made of 4mm polycarbonate, 18" tall, and keep cold air off plants while still allowing hot air to escape. Our greenhouse supplier had sheets of 4mm polycarbonate that he sold us for a steep discount. We have also seen short walls that use greenhouse plastic in place of the polycarbonate.

Drop-down sides are another more complex solution to the same problem, but they would not work with our shape of greenhouse (straight sides are best). I suggest talking with your supplier if you are considering drop-down curtains.

The greenhouse pictured on the previous page shows the system of shutters on the front. The photo above shows the back of the greenhouse with two exhaust fans that kick on at different temperatures. Photos by Ben Hartman.

Problem: Weeds — and ice — creeping in under baseboards!

Solution: Insulation and landscape fabric.

This spring we spent many hours weeding the inside perimeter of our greenhouses. No matter how many we pulled, by the next week a new crop was trying to sneak in. When we added up the number of hours we were spending on the task, we realized it would be worth it to block out the weeds with landscape fabric.

Another problem we had encountered in the winter was that the ground in our outer beds was often too wet to plant (we have heavy clay), or even frozen.

The author's greenhouse has short sidewalls to protect plants at ground level from cold wind. It also has rigid foam insulation board, covered with weed barrier and wood chips, to keep the soil from freezing in winter, and prevent weeds in summer.

Our solution to these problems was to first improve our grade. Then we laid 1" thick by 24" wide pink foam insulation around the perimeter of our greenhouses. (We have heard of this practice in Scandinavia, though we have not actually seen in done.) Next, we covered the insulation with landscape fabric. We attached the fabric to the baseboards with lathe strips we cut from exterior-grade lumber. Finally, we covered the fabric with wood chips.

The insulation effectively extends the roof canopy of our greenhouse out 24" from the edge. So it both sheds water and (we expect) will help keep the greenhouse soil closest to the edge from freezing. It has already helped our perimeter beds stay drier. And we have saved many hours of weeding, about which no one can complain.

Ben Hartman and Rachel Hershberger own Clay Bottom Farm near Goshen, Indiana. For more information and videos about their greenhouse systems, visit www.claybottomfarm.com/grower-resources/

Next generation: movable houses

By Lynn Byczynski

Eliot Coleman broadforks the soil in preparation for planting the next crop under a movable high tunnel. Photos courtesy of Bob Rimol, Rimol Greenhouse Systems.

As high tunnel production has gained popularity, an increasing number of growers are moving on to what could be considered the next level of complexity — movable high tunnels. Eliot Coleman has been promoting movable hoophouses for more than 15 years, but they have been slow to catch on, possibly because they required more engineering and building skills than most people possess.

In the past few years, however, there has been a proliferation of mobile hoophouses. Several companies are offering ready-made movable tunnels. Some of the top hoophouse researchers at universities have purchased movable tunnels. And new solutions have been developed for anchoring tunnels more securely, alleviating the biggest fear for growers in windy areas.

Here's the latest news on production and products in the ever-evolving science of hoophouses.

Increased productivity

Most high tunnel growers have one major complaint: there's never enough space to grow everything they want. Growers with movable high tunnels, however, think they have solved that problem. With careful scheduling, they can use the valuable high tunnel space when it's most needed by a crop, then move the tunnel to protect the next crop.

Eliot Coleman, who continues to refine his pioneering work with mobile tunnels, says that he pays for one greenhouse, but in effect has two. "If you add up all the months those hoophouse crops take from start to finish, it's 24 months."

The number of crops that can use the same hoophouse depends on the number of separate plots that can be covered. At its simplest, a tunnel moves back and forth over two areas. For example, spinach can be grown in winter and, when it's able to survive outside in early spring, the tunnel can be moved and tomatoes planted beneath it. Two movable tunnels allow for a four-year crop rotation.

A longer run covering more plots brings more complexity and expanded potential for crop rotation. In his book *Winter Harvest Handbook*, Coleman outlines a crop sequence involving four plots and five moves of the tunnel per year. He concludes that a 1,000 square foot hoophouse is able to produce 1,000 square feet **each** of onions/scallions, carrots/beets, zucchini/beans, early tomatoes, late fall cucumbers, and late lettuce.

"That is making very good use of the slight increase in capital investment to purchase a movable house compared to a static house," he writes.

The key to maximizing yield is to schedule crops in progression based on their cold-hardiness. A tunnel that covers spinach in winter can be moved aside in early spring; the spinach will do fine with light frosts outside and, under cover, a new crop of carrots can be planted. A month later, the carrots will be fine outside, so the tunnel can be moved again and early tomatoes planted. And so on, with something growing under cover year-round.

Pete Johnson of Pete's Greens in Craftsbury, Vermont, grows tomatoes, peppers, and eggplants in his movable hoophouses in summer. He needs to start fall crops such as spinach, mache and claytonia in September in order to have enough growth before the short days of winter arrive. With stationary tunnels, he would have to pull out the tomatoes two months before they were done producing in order to get the winter crops started.

But with the movable tunnels, he starts the winter crops outside, adjacent to the tomato tunnels. The greens are hardy to about 24°F, so when the forecast is for temperatures below that, he moves the tunnel onto them. The tomatoes would have been killed by temperatures that low, even in the tunnels, so no time is lost on either crop. The only down side, he said, is that the soil is pretty wet by the time he moves the tunnels, which causes a lot of moisture in the tunnel all winter and limits the types of crops he can grow there.

Dr. Ed Page of Colorado State University Extension in Montrose has a SARE grant to research scheduling with a movable tunnel over spring greens followed by summer crops followed by fall raspberries. Even with the movable tunnels, he finds, growers have to choose the most profitable crop because of overlaps in their schedules.

"I'm trying to demonstrate that there might be a way to use the hoophouse to get more production hours out of it," he said. "So many of these hoophouses will sit empty for six months. If you can use that one hoophouse to cover three, four, or five different crops...I think it makes a lot of sense for people who have intensive crops and small areas to work with."

Disease prevention

Alex and Betsy Hitt of Peregrine Farms in North Carolina were early adopters of mobile greenhouses, putting up their first in 1997. For them, the issue was not extending the season in fall — they want their winters off. Rather, as organic growers, they were concerned about the potential buildup of soilborne diseases that might occur in a stationary hoophouse where the soil never freezes.

"We're sliding them back and forth so the soil isn't always covered up," Betsy Hitt said. "We think the main advantage is better soil health; we have never had a problem with diseases of a soilborne nature."

The Hitts purchased three 16' x 96' high tunnels, but divided them into six sections of 16' x 48'. Half are devoted to cut flowers and half to vegetables. Betsy notes that using movable tunnels for flowers is completely different from using them for vegetables. Cold hardiness is not the most important factor in determining when the tunnel is over the flowers. "You generally want flowers covered when they are flowering," she said, to protect the blossoms from wind and rain damage. She grows many flowers under Haygrove tunnels, which are not movable but are easily uncovered, which provides the same soil-cleansing effect as movable tunnels.

Dr. John Biernbaum at Michigan State University got a mobile tunnel and plans to put it up this year. "I think anybody looking at tunnels now should be looking at them (movables)," he said. Besides helping with disease prevention, the mobile tunnel also allows for cover cropping for soil fertility.

Structural improvements

Coleman's earliest models were built on skids and required a tractor to drag them to a new position. Now he has several new tunnels that are much more user-friendly. The current crop of hoophouses are from Four Season Tools and Rimol Greenhouse Systems. Both companies manufacture their own movable structures. Rimol's "Rolling Thunder" model has a round pipe that serves as a rail, with wheels attached to the hoops. Four Season Tools has a "V-track" design, also with wheels on the hoops. "If I had to, I could move it myself," Coleman said of his Four Season Tools mobile V-track tunnel.

Greg Garbos of Four Season Tools says that a 30' x 48' movable tunnel ranges from about $7,000 for a skid-based system to $8,500 for a V-track system. In general, he says, a movable tunnel will cost 30% to 50% more than the same size stationary tunnel.

Top photo shows Four Season Tool's V-track hoophouse, with a detail above of the track and wheel system. An added benefit of the V-track system is that the wide piece of metal prevents weeds.

Rimol's Rolling Thunder model costs about $10,000 for a 22' x 48' with polycarbonate panels for endwalls, about $1,000 less for poly endwalls. Prices fluctuate with the price of steel, and Garbos noted that the price of some steel components has increased 38% since last fall.

Many farmers design and build their own movable tunnels. Some start with a manufactured high tunnel kit and modify it to make it movable. That's what Pete Johnson did with 35' x 200' Harnois greenhouses he built two years ago. His system consists of 4" x 4" angle iron for skids. The hoops fit into brackets, designed to bolt a greenhouse onto concrete, that are attached to the skid. The setup requires two tractors with bucket loaders and winches to pull the tunnels when it's time to move them. He tried making a video of moving the tunnels, but joked that it was too boring to watch because they move so slowly.

"We can move all four of these with two people in about six hours," Johnson said, "so it's not a big deal."

The Rimol hoophouse pictured on page 8 has wheels bolted to hoops that run on a round galvanized metal pipe. Photo courtesy of Rimol Greenhouse Systems.

Anchoring systems

Anchoring the structure is probably the most critical factor in building any hoophouse, but especially a movable one.

"Movable greenhouses are like parachutes staked to the ground," said Ben Hartman of Clay Bottom Farm in Goshen, Indiana. "Our experience has taught us to not underestimate the power of the wind to set these structures sailing.

"In May, 2009, we purchased a used 30' x 150' greenhouse at auction. Our plan was to divide the structure into three separate movable hoophouses. We built the first hoophouse, a 30' x 45' structure, in August-September 2009, and planted it with spinach and salad greens on September 15. We harvested greens about five weeks later. Then a strong wind lifted the hoophouse and rammed it into the roof of our barn, nearly 150' away and more than 10' in the air. We joke that our experiment was a success — it moved!

"In reality, we're glad that no one was hurt and that damage to the barn was limited to minor roof repair. Most of the hoophouse, however, was unsalvageable, except for the galvanized 2" x 3" steel skids on which it was built. We're saving those for a future, better-engineered, movable hoophouse."

Greg Garbos says that many people mistakenly think a tunnel will be safe in a storm if the doors are closed and the sides rolled down. "There are essentially two concerns with wind affecting a mobile greenhouse — it will find its way underneath a structure and cause lift, similar to a kite, or wind can move over the structure causing lift, like an airplane wing," he said. As a result, closing the house is important when a storm is imminent, but the structure also needs to be well anchored.

Rimol and Four Season Tools have different anchoring systems for their mobile hoophouses. That's an important variable, and prospective customers should review the details before choosing one over another.

Although it's not recommended to put wheels on a hoophouse that was designed to be fixed, some people have adapted anchoring systems for fixed structures. Growers with experience advise that moving the house will be a lot easier if the system for unbolting the house from the anchors is accessible and simple.

Farm Tek sells a ground anchor assembly that bolts to the greenhouse hoop at one end and to the earth anchor at the other end. Pete Johnson

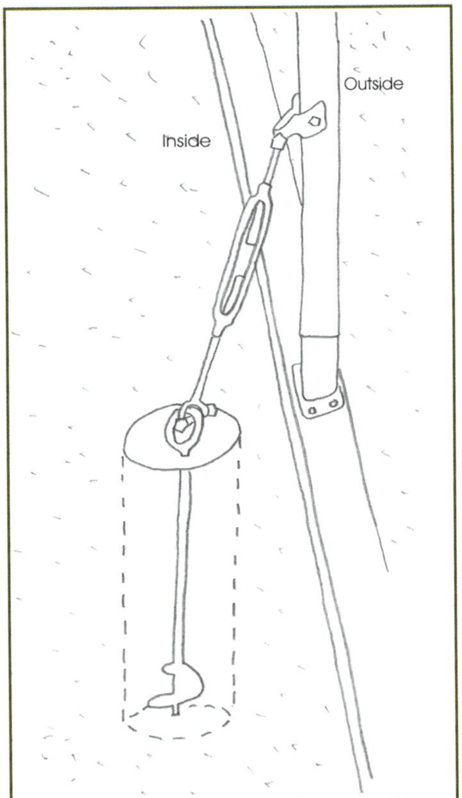

A ground anchor system for movable greenhouses: To anchor the greenhouse to the posts, set a steel earth anchor into wet concrete, 12" in diameter and 3' deep. After the cement cures, heavy duty steel turnbuckles can be used to connect the top loop of the earth anchor to a pipe banding clamp on the greenhouse rafter leg. If the anchor is buried on the inside of the greenhouse it will not interfere with greenhouse plastic. The system should be used on all four corners, and could also be used along the sides. Diagram by Ben Hartman, Clay Bottom Farm.

anchors his 200-foot-long hoophouses with a variation. He put bolts in the hoops about 4 feet high and drove augur-style earth anchors from Farm Tek into the ground inside the house and below each bolt. Come-alongs are used to connect the bolts in the hoops to the earth anchors. The ratchet action of the come-alongs make it easy to tighten and release the cables.

How many anchors are needed to hold down a house depends on several factors. The weight of the hoophouse itself is often used as the guide, but Garbos thinks that's not the most important factor. Other variables include soil type, which affects the holding capacity of the anchors, and climate, including the frost line in winter. In the case of Ben Hartman's destroyed hoophouse, he said "at the time, we thought we were erring on the side of over-engineering." That house was anchored three ways:

- T-posts were driven deep into the ground and tied securely at all four corners.
- 2" x 4" lumber, which framed the doors, were bolted onto a 2" x 6" cemented in the ground.
- 3' long, 5/8" rebar stakes (welded locally) were driven at opposite angles through steel hoops every 10 feet along the side of the hoophouse.

"All three systems failed," Hartman said. "Posts and stakes pulled up; the cement posts stayed in the ground, but the bolts ripped through the 2" x 6" lumber."

Garbos cautions against using any generic formula in deciding how to anchor a movable hoophouse. He advises growers to purchase a hoophouse that was designed and manufactured to be mobile. Most fixed structures do not have the internal bracing necessary to make them mobile, he says. But for those who are going to try to retrofit their tunnels to make them movable, he suggests talking to local building code officials about requirements for anchoring a mobile home.

Local requirements for anchoring a permanent greenhouse can also be helpful. But be aware that in some jurisdictions, greenhouses are taxable but hoophouses are not. If concrete anchors are used, the house may move from the category of temporary agricultural structure that is non-taxable and be considered a permanent structure that goes on the tax rolls.

Top photo shows Clay Bottom Farm's movable hoophouse a week before it blew away. Below, it landed on the barn roof, and was not salvageable. The hoophouse had a quick-to-build endwall design, which allows for maximum ventilation and for the hoophouse to be moved over plants. The endwall, however, lacks back bracing so the door is loose and catches wind—this might have contributed to the disaster. Photos by Ben Hartman.

How to get snow off the hoophouse

Although most hoophouses can withstand some snow load, if you've got too much snow, you need to get it off. You can attack it from the inside first, but if that doesn't work, you'll have to go outside.

From the inside: Using a broom, sweep down the sides of the greenhouse poly to dislodge the snow. You've got to start at the bottom and work your way up as the snow cascades down. A wide push broom will remove more snow per sweep, but it gets heavy holding it above your head for any length of time. You may prefer a lightweight kitchen broom. This sweeping usually works just fine when the snow is light and dry. But if it's heavy and wet and won't slide down, you need to go outside with a partner.

Outside: Get a length of rope long enough to extend across the top of the hoophouse. Tie knots in it every couple of feet, and throw the end across the hoophouse to your partner. Then just pull the rope back and forth between you; the knots help dislodge the snow. This is when it gets really fun, as sheets of snow cascade down into your boots!

A farm-built movable tunnel

The author's 48 x 22' high tunnel rolls on 200 feet of track. The long track provides four plots, giving him great flexibility in scheduling crops that need the protection of a high tunnel. All photos courtesy of Phil Norris.

By Phil Norris

Here in Maine, the short growing season makes some sort of season extension desirable. Unheated high tunnels have proven themselves to be profitable and movable high tunnels have been around long enough for the advantages to be evident. Several companies are putting out well-engineered rolling greenhouses and the farmer now has some choices.

I wanted a movable high tunnel so that I could get away from the insects that plague stationary greenhouses. I wanted this high tunnel to roll easily so that it could be moved several times each season. I wanted plenty of clearance so that the high tunnel could roll over established crops. And I wanted it to vent entirely through the endwalls.

I had originally planned for a 60' X 22' house. I have to admit that in designing this movable high tunnel I leaned heavily on the advice of my neighbor down the road, Eliot Coleman, who has built many rolling greenhouses. It was he who convinced me that a 60' high tunnel wouldn't vent through the end walls, that there would be a dead spot in the middle. He suggested that a 48' X 22' house would. He was right.

Building the frame of the high tunnel. Notice the steel cross bar at waist height on both endwalls, which provides structural stability when the tunnel is moved.

Designing the endwall was the biggest challenge. Most manufacturers provide the basic frame including the bows and cross members but they leave the endwall up to you. I drew various endwall possibilities but the entrance always seemed to be the structural sticking point. It seemed to me that the doorway destroyed the integrity of the wall. And I sure as heck didn't want to have to duck under a beam to enter the house. The answer: a sidewall entrance. That meant I could use a 22' long piece of steel at waist height on both endwalls. That would be integrity. You hardly ever see entrances on the sidewalls of greenhouses. Could it be that difficult?

I built the house in the fall and got it covered with poly just before the cold weather set in. At first the sidewall door was a flap of poly sealed around the edges with wiggle wire. It worked that first winter and kept the snow out but it was cumbersome. It wasn't elegant. In the spring I bent some steel tubing to the same shape as the sidewall and welded up a door frame then screwed on a piece of polycarbonate. I hung this four foot wide door with standard sliding barn door hardware which consists of an 8' hollow galvanized steel track with two wheeled "trucks" that ride inside it. The door is hung from the trucks. The bottom of the door is free and is held down by gravity.

Because he didn't want to use the endwall for access, the author designed a sliding door on the side of the high tunnel.

The roll-up endwall is fastened with wiggle wire to a piece of double wire lock positioned 42" above the soil. This is so that when the poly is rolled up it will be well above the lower beam, giving 32 honest inches of clearance. The rolled up poly is secured in the "up" position by slipping a rope loop over the end of the roller. It does take two people to roll up the endwalls and it takes two people to move the house.

Sprinklers hang above crops in the high tunnel. One of the unexpected benefits of the design is the ability to use the tunnel as a movable sprinkler system.

I covered the west endwall with poly using a scissors-type closure. Simple and cheap and it has worked out fine. The east wall is covered with rigid polycarbonate because the greenhouse is exposed to battering east winds and I wanted a strong wall in that direction. It features removable panels for the warmer seasons.

The house rolls on two 2" galvanized pipes. My site is slightly curved so I laid out the pipes with a gentle curve and then built the house on top of them. The whole greenhouse is uniformly curved. The house rolls more easily than any other movable greenhouse I have ever seen. Two people can push it by leaning on the corners. No tractors, no winches. We have to tie it down in the summer just so a breeze won't send it down the track. Pre-tensioning each bow with a come-along and then locking it in at exactly 22' before erecting it took some extra time but has probably contributed to ease of rolling. The strong endwalls with an uninterrupted beam 32" above the soil and diagonal braces to keep the wheels from splaying out also probably contributed.

The house is anchored by a 3' ground post driven into the clay just off each corner. The corner chains lead from these ground posts to eyebolts on the house. Six additional ground posts are used inside the house, three along each side, giving a total of ten ground posts for the winter position of the house. In the summer, when the winds are mild, a quarter-inch piece of nylon rope at each corner has been more than adequate. If a hurricane is expected we will roll the house to the winter position and secure it down.

I wanted some sort of simple automatic vents, not necessarily capable of venting the whole house, but to exhaust some of the heat in case a cloudy day suddenly turned sunny. The obvious place to put the vents was right at the peak. I used hinged triangular shaped vents made of polycarbonate on a wooden frame attached to a couple of nonelectric vent openers. These vent openers are sealed hydraulic tubes that are designed to lift a heavy vent. In this horizontal application they work effortlessly and have worked flawlessly. They start to open the vents at about 75°F and are wide open at 85°F. I also hung a wireless thermometer in the middle of the house which beams a signal to my office. It monitors and graphs the temperature and humidity in the high tunnel and during a Maine summer, with both endwalls fully vented, the temperature in there never went over 99°F. The peppers loved it.

Another unexpected use of the high tunnel has been as a rolling sprinkler. I installed an overhead system with 32 sprinklers which attaches with a garden hose to any one of four water spigots. It only takes a minute to roll the house onto any one of the four plots, hook up the hose, and water a thirsty crop. It takes two hours to provide an inch of rain, after which the house can be rolled back to its starting position.

The tunnel is ventilated with a scissors-type closure and triangular vent at the peak. The exterior view below shows the vents fully opened.

At our farm we still have to work out the best planting dates and rotations for this four-plot system. We envision that the tunnel will be tied down and zipped up securely for a winter crop of greens. Then in the spring it will be rolled over a bed of strawberries. When the strawberries are well underway the house will be rolled onto a summer crop, such as peppers and melons. And then in October, the house will be rolled onto a fall crop of kale, Asian greens, lettuce, and spinach. One plot will always be in cover crops such as peas, vetch and oats.

At the end of our first growing season with this high tunnel, we are amazed at how much produce we have harvested from these four 48' plots. The high tunnel hasn't paid for itself in one year and we didn't expect it to. But it is well on its way.

Phil Norris is the owner of Clayfield Farm in East Blue Hill, Maine. He can be contacted at phil@clayfield-farm.net.

Caterpillar tunnels

An inexpensive variation on the hoophouse theme

By Lynn Byczynski

Among the many season extension structures in use on vegetable farms today, one of the least expensive is the caterpillar tunnel. Caterpillars — so named because of their segmented appearance — are constructed of pvc pipe, rebar, and rope. They can be up to 300 feet long, and don't require flat ground. Most are 5 to 6 feet tall at the center of the hoop, so the grower can walk into them from any point along their width (although it's not as comfortable as working in a high tunnel).

Caterpillars are easy to build and move. They are inexpensive compared to permanent hoophouses and most of the materials can be found locally. They provide protection but, at the same time, excellent air flow.

Caterpillars are used for many crops and many purposes. They can be covered with poly early and late in the season to provide a warm environment. They can be covered with shadecloth in summer to grow lettuce, cut flowers, or other cool-loving plants. When covered with row cover (e.g. Agribon) or insect mesh, they can be used in summer to exclude pests.

This article will focus on growing summer crops, including cucumbers, tomatoes, eggplants, and flowers in a caterpillar tunnel.

Building a caterpillar

The hoops for a caterpillar tunnel can be galvanized steel, electrical conduit or pvc. They can be pushed directly into the soil, or placed over rebar ground stakes, with about a foot in the ground in either case. Hoops should be spaced 6-10 feet apart, depending on wind exposure. A rope is then tied from hoop to hoop to form a purlin that runs the length of the tunnel. The rope should be tied off to stakes at both ends of the tunnel. The covering is then pulled over the tunnel and held to the ground with pegs, sand bags, or rocks. Stakes are pounded into the ground between hoops on both sides of the tunnel. Ropes are tied across the top of the tunnel to the stakes to hold the covering in place.

Anne and Eric Nordell of Beech Grove Farm in Trout Run, Pennsylvania, use caterpillars on their farm. Here is how they build them:

They set 30-inch pieces of 5/8-inch rebar 6 feet apart and 18 inches deep (12 inches of rebar sticking out of the ground) in two rows 10.5 feet apart. For the hoops, they use 20-foot lengths of Schedule 40 1-inch pvc pipe. They slide the pvc pipe over the rebar stakes, forming flexible

Eric and Anne Nordell built a 102' long caterpillar tunnel to grow early cucumbers at their farm in Pennsylvania. They covered it with insect mesh and greenhouse poly. Photo by Anne Nordell.

hoops. For the ridge purlin, they use clothesline and loop it around the top of each hoop and tie it off to 36-inch rebar posts that are set about 8 feet past each end of the tunnel. They attach a separate rope from the end hoop to the stake to add stability.

Cucumbers and tomatoes

The Nordells have a short season at their farm in the mountaints. They have found the caterpillar tunnel to be a cost-effective way to get growing earlier and later, and to protect crops from insects and disease. They use a caterpillar for lettuce, spinach, salad mix, cucumbers, tomatoes, basil, chard and kale.

For the past three years, they have participated in a trial of organic control of cucurbit pests sponsored by the Pennsylvania Association for Sustainable Agriculture (PASA). The Nordells focused on early cucumber production using Enviromesh anti-insect screen and 6-mil greenhouse poly on a caterpillar, and compared it to their standard practice of growing cucumbers under floating row cover outside.

For pollination, they removed the floating row cover when the plants flowered in the field; inside the caterpillar, they placed bumblebee hives.

Yields were 78 percent higher in the caterpillar than the field. The field cucumbers had a 28 percent cull rate, all because of damage by cucumber beetles, compared to 20 percent culls in the caterpillar, all caused by misshapen fruit. Some cucumber beetles did manage to sneak into the caterpillar, but in general it did a good job of excluding them. The caterpillar also accelerated the first harvest of cucumbers by nine days compared to those under floating row cover.

The caterpillar also saved the Nordell's tomato crop last year. In the field, the first tomato harvest is normally the third week of August. But late blight, which was rampant all over the East in 2009, killed their field tomatoes in mid-July, before they started producing. But the tomatoes in caterpillars started bearing in early July and kept on going until late August, when they finally succumbed to late blight. In other words, the caterpillar provided six weeks of earliness and six weeks of late blight protection.

The insect screening, while obviously essential for pest control, also seemed to improve the stability of the caterpillar tunnel, Eric said. Whereas spunbonded row cover such as Agribon is susceptible to being torn off by wind, the insect screening lets air move through the tunnel, resulting in less damage to the structure and better air circulation for the crops.

Enviromesh, also known as Proteknet, insect screen is available from Purple Mountain Organics (see ad on page 4 for contact information) In the U.S. and from Dubois Agrinovation in Canada (they will ship to the U.S. See www.duboisag.com). Other brands of greenhouse insect screening are available from G&M Agricultural Supply (www.gmagsupply.com).

Cut flowers

Growing cut flowers in caterpillars has resulted in longer stems and higher quality blooms for Jan and Ted Blomgren of Windflower Farm in Easton, NY. They started using the tunnels with row cover to exclude leafhoppers, which carry aster yellows, a disease that damages many flowers in the Aster family. Not only were they able to grow China asters without leafhopper damage, they also got 3-foot stems. Since then, they have grown many kinds of flowers in caterpillars, including bells of Ireland, godetia, lisianthus, snapdragons, and stock. Now they cover the flower caterpillars with 3-4 mil greenhouse film, not row cover, and grow asters under row cover in low tunnels. The Blomgrens have about 12,000 square feet of the tunnels on their farm for production of vegetables as well as flowers. Many more details are included in the manual *High Tunnels: Using Low Cost Technology to Increase Yields, Improve Quality, and Extend the Growing Season.* Ted Blomgren is co-author with TracyFrisch and Steve Moore of this excellent resource, which was published by the University of Vermont Center for Sustainable Agriculture. It covers many kinds of season extension structures, from low tunnels to Haygrove multi-bay tunnels. See Resources on the next page for information about where to obtain the manual.

A caterpillar tunnel is used to protect tomatoes at Johnny's Selected Seeds research farm in Maine. The tunnel was built onsite using Johnny's Quick Hoops High Tunnel Bender to make the hoops. Photo courtesy of Johnny's.

Part 3 Resources

Information resources

Books mentioned in *The Hoophouse Handbook* are available from *Growing for Market*, by phone (785-748-0605) or online at www.growingformarket.com. Prices and more details are included in the book list on pages 109-110.

High Tunnels: Using Low-Cost Technology to Increase Yields, Improve Quality and Extend the Season is a 74-page booklet published with SARE funding by the University of Vermont Sustainable Agriculture program. It is available for free download at http://www.uvm.edu/~susagctr/resources/HighTunnels.pdf

Vegetable Production Budgets for a High Tunnel - Linda Naeve Iowa State University Extension. 8 pages, May 2012; https://store.extension.iastate.edu/Product/Vegetable-Production-Budgets-for-a-High-Tunnel

University of Kentucky *High Tunnel Tomatoes* is available from www.uky.edu/Ag/CCD/introsheets/hightunneltomatoes.pdf

There is a wealth of information on the web about hoophouse construction and growing. Do an internet search and you'll find countless pages and videos. A more helpful way to search is to name your state + high tunnels. Here are several resources that will help narrow the search and provide additional links to high-quality information about commercial high tunnel production:

www.hightunnels.org — This site is the result of a collaboration among Extension specialists and grower-cooperators in Kansas, Missouri and Nebraska. Funded in part by USDA, the high tunnel project covers vegetable and cut flower production in the Midwest and serves as a clearinghouse for high tunnel information from other states. The website also maintains a calendar of events, including conferences, workshops, and field days.

Federal funds available to buy high tunnels

Federal funding is available for farmers to purchase high tunnels. The Seasonal High Tunnel Initiative is administered by USDA's Natural Resources Conservation Service as part of its EQIP program. Established in 2009 as a pilot program, the High Tunnel Initiative became an established conservation practice standard in 2014.

In the first four years, NRCS funded more than 13,000 high tunnels nationwide at a cost of more than \$61 million, an average of about \$4,700 per high tunnel.

The program is offered in every state, plus Puerto Rico and the Pacific Basin. The amount of funding, application deadlines, and other details are determined by each state, and there is quite a bit of variation. To learn about the program in your state, contact your local NRCS office. A list of NRCS service centers is available here: http://offices.sc.egov.usda.gov/locator/app?agency=nrcs

The original program set a maximum size of 2,178 square feet of high tunnel that could be funded but that cap was lifted iin 2014. Now states can set limits either on the size of the tunnel or the amount they will pay per square foot, to insure the money gets spread around equitably.

The top 5 states where high tunnels were built with federal funding are Missouri, Alaska, Michigan, Wisconsin, and Minnesota. It's not surprising that high tunnels are popular in those states with cold winters, but southern growers are taking advantage of the program, too, with the next four-highest states being Kentucky, Alabama, Georgia, and North Carolina. Beginning farmers have been especially keen to build high tunnels; they accounted for 51% of all the contracts with NRCS in the first four years of the program.

The National Sustainable Agriculture Coalition is keeping tabs on the Seasonal High Tunnel Initiative. For the latest news, check http://sustainableagriculture.net/blog/high-tunnel-update/

www.attra.org — ATTRA, the National Sustainable Agriculture Information Service, offers dozens of publications about market farming, including several on greenhouse and hoophouse production. Most publications can be downloaded for free, or you can order printed copies for a fee. You can also call toll free, the English-language ATTRA at 800-346-9140 or the Spanish-language ATTRA hotline at 800-411-3222.

http://plasticulture.cas.psu.edu — Penn State's Center for Plasticulture posts results of high tunnel vegetable research on this site.

www.noble.org — The Noble Foundation continues to research hoophouse production for the South, and posts reports here periodically.

A video series on high tunnel construction is available at www.sare.org/Learning-Center/Multimedia/North-Central-SARE-Multimedia/Kansas-State-High-Tunnel-Video-Series

Growing Under Cover is a publication of the Kansas Rural Center, available as a free download from www.kansasruralcenter.org/growing-under-cover

www.hort.cornell.edu/hightunnel — Cornell University's home page for publications about high tunnel growing.

http://www.uky.edu/Ag/CCD/hightunnel.pdf Plans from the University of Kentucky for a homemade PVC hoophouse.

Hoophouse manufacturers

Agra Tech; 925-432-3399;
www.agratech.com

Atlas Greenhouse Systems; 800-346-9902;
www.atlasgreenhouse.com

BWI Companies, 903-838-8561;
www.bwicompanies.com

Conley's Greenhouse Manufacturing & Sales,
800-377-8441; www.conleys.com

DripWorks, 800-522-3747;
www.dripworks.com

Four Season Tools, 816-444-7330;
www.smallfarmtools.com

G&M Ag Supply, 800-901-0096;
www.gmagsupply.com

Golden Pacific Structures; 803-348-0161;
www.gpstructures.com

Griffin Greenhouse & Nursery Supplies;
978-851-4346; www.griffins.com

Growers Solution, 866-928-3390;
www.growerssolution.com

Growers' Supply by Farm Tek, 800- 476-9715;
www.GrowersSupply.com/ADGFM

Harnois C.P., 450-756-1041;
www.harnois.com

Haygrove; 717-492-4955;
www.haygrove.com

Hummert International, 800-325-3055;
www.hummert.com

Ledgewood Farm, 603-476-8829;
www.ledgewoodfarm.com

Ludy's Greenhouse Manufacturing,
800-255-5839; www.ludy.com

Morgan County Seeds, 573-378-2655;
www.morgancountyseeds.com

Poly-Tex, 800-852-3443;
www.poly-tex.com

Rimol Greenhouse Systems; 877-746-6544;
www.rimolgreenhouses.com

Stuppy Greenhouse Manufacturing,
800-733-5025; www.stuppy.com

Tunnel Vision Hoops; 330-283-5717;
www.tunnelvisionhoops.com

The Best Books about Market Farming

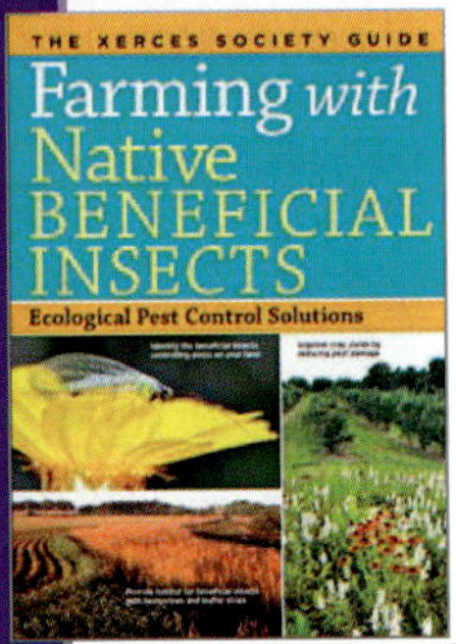

• **NEW!** ***Farming with Native Beneficial Insects***. This beautiful and inspiring book by the Xerces Society shows how to create a farm or garden habitat that will attract native beneficials to protect your crops from damage without pesticides. Great photos and detailed directions for attracting and conserving these important allies on a sustainable farm. $25 **Member price: $20**

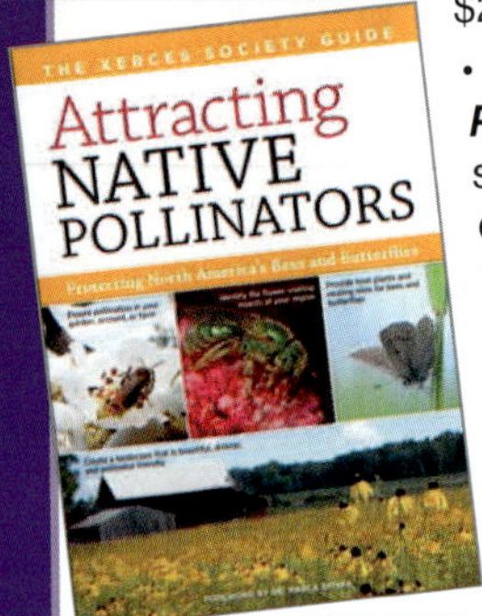

• **NEW!** ***Attracting Native Pollinators.*** The Xerces Society shows you how to provide flowering habitat and nesting sites for pollinators. With native pollinators responsible for pollinating 70% of all flowering plants, including many vegetables and fruits, ensuring their survival is critical to market farmers. $30 **Member price: $24**

• ***Sustainable Market Farming:*** *Intensive Vegetable Production on a Few Acres.* GFM Contributing Editor Pam Dawling provides the kind of details you need to successfully grow vegetables on a few acres: varieties, scheduling, cultivation, harvest, and postharvest for every vegetable are covered, plus chapters on sustainable practices. $35 **Member price: $28**

• ***Sustainable Vegetable Production from Start-Up to Market*** by Vern Grubinger. A great introduction to market farming on the livelihood scale — excellent for small growers interested in scaling up. $38 **Member price: $30.40**

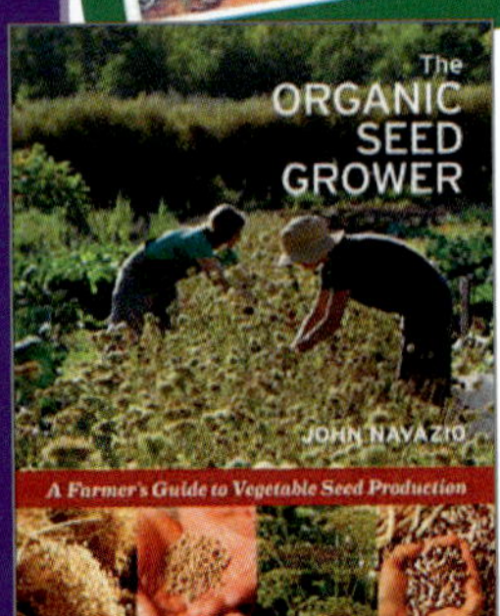

• ***The Organic Seed Grower***. Dr. John Navazio has written a clear, comprehensive guide for growers who want to grow seed as a business or for their own use. He advises on each type of vegetable, including selecting the best plants for your conditions. $50 **Member price: $40**

• ***Crop Rotation on Organic Farms***. Provides an in-depth review of crop rotation and its many applications, such as improving soil quality and health, and managing pests, weeds, and diseases. $24 **Member price: $19.20**

• ***Winter Harvest Handbook.*** Coleman's work for the past decade has focused on extending the season with unheated hoophouses. In this book, he shares everything he has learned about growing and harvesting in fall, winter, and early spring. Essential reading for new hoophouse owners. $30 **Member price: $24**

• ***The Organic Farmer's Business Handbook*** by Richard Wiswall. An experienced, lifelong farmer shows how to identify profitable crops, increase efficiency, and make more money on your market farm. Includes CD with self-calculating spreadsheets for creating crop budgets. $35 **Member price: $28**

• ***The New Organic Grower*** by Eliot Coleman. *A Master's Manual of Tools and Techniques for the Home and Market Gardener.* Everyone reads this book when starting out in market gardening! $25 **Member price: $20**

• ***Growing and Selling Fresh Cut Herbs*** by Sandie Shores. The best how-to book for commercial production of culinary herbs. $35 **Member price: $28**

• ***Growing Great Garlic*** by Ron Engeland. The first garlic book written for organic gardeners and small-scale farmers. $17 **Member price: $13.60**

• ***Microgreens*** by Eric Franks and Jasmine Richardson of Microcosm Microgreens in Big Sur, California. Provides detailed advice on producing microgreens for market or personal use. Illustrated with beautiful photographs and includes numerous recipes. $20 **Member price: $16**

GFM subscribers get 20% off all books!

Index